Recent Advances in the Treatment of Neurodegenerative Disorders

Edited by

Sachchida Nand Rai
Centre of Biotechnology
University of Allahabad
India

Recent Advances in the Treatment of Neurodegenerative Disorders

Editor: Sachchida Nand Rai

ISBN (Online): 978-1-68108-772-6

ISBN (Print): 978-1-68108-773-3

ISBN (Paperback): 978-1-68108-774-0

need for a court order if at any point you breach any terms of this License Agreement. In no event will any delay or failure by Bentham Science Publishers in enforcing your compliance with this License Agreement constitute a waiver of any of its rights.

3. You acknowledge that you have read this License Agreement, and agree to be bound by its terms and conditions. To the extent that any other terms and conditions presented on any website of Bentham Science Publishers conflict with, or are inconsistent with, the terms and conditions set out in this License Agreement, you acknowledge that the terms and conditions set out in this License Agreement shall prevail.

Bentham Science Publishers Ltd.
Executive Suite Y - 2
PO Box 7917, Saif Zone
Sharjah, U.A.E.
Email: subscriptions@benthamscience.net

BENTHAM SCIENCE

CONTENTS

FOREWORD

A limited boom in neurodegenerative diseases (NDDs) and their treatments attract scientists all around the world. However, the conventional medications of NDDS are not sufficient to provide a better cure for patients. Surgical alternatives also have minimal efficacy with different kinds of side effects. Likewise, advancement in alternative treatment for different NDDs is essential nowadays. Chemically mediated therapy also have a minimal impact, which is associated with significant side-effects. The blood-brain barrier is the biggest culprit to minimize the response of different drugs. Nanoparticles and nanoformulation-based treatment can overcome the BBB problem and improve the efficacy of treatments. Therefore, eating habit also has a significant impact on the management of NDDs. The herbs and plant extracts serve as a better alternative with minimal side-effects. However, edible mushrooms can treat various kinds of NDDs, like Parkinson's disease (PD) and Alzheimer's (AD).The gut biome modulates the therapeutic efficacy in most common NDDs like PD and AD. Polyphenols also show the maximum impact on the treatment of NDDs. Moreover, animal models play a vital role in the standardization of drugs for the treatment of NDDs. Different yoga postures and techniques have a beneficial impact on the management of different neuronal diseases. As mentioned earlier, these options are the most advanced alternatives for treating diverse kinds of brain disorders. In this book, the editor has included all the above-mentioned recent advancements for the medication of neurological disorders. This book attracts the interest of researchers and scientists to explore the current treatment option in their researches. It draws both basic and clinical kinds of research to utilize a different alternative option that will be very efficient in treating NDDs. The literature in this book will be significantly crucial for the academicians, molecular biologists, graduates, and undergraduate students engaged in basic and clinical research. The mentioned distinct tools and techniques in this book can unravel the problem of different NDDs. However, we believed that the information that will be gainedby reading the chapters includedin this book, edited by Dr. Sachchida Nand Rai. The later exchange on every topic serves as an essential and valuable tool to understand the different and more advanced alternative treatment options for different Neurological disorders.

Emanuel Vamanu
University of Agricultural Sciences and
Veterinary Medicine
Bucharest
Romania

PREFACE

The central nervous system (CNS) is the most vital component of our body, regulating various kinds of daily activities that are essential for our life processes. Keeping the balance between body and brain and maintaining the homeostasis of CNS is one of the main focuses of researchers nowadays.

Neurodegenerative diseases (NDDs) arise as a result of progressive degeneration of neurons in the CNS. Researchers have tried various effective treatments that prevent this progressive neurodegeneration of neurons within the CNS. Parkinson's disease (PD), Alzheimer's disease (AD), Multiple sclerosis (MS), *etc.,*are some of the most common NDDs. Conventional treatment has limited success in the treatment of NDDs. The primary aim of this book is to provide an audience worldwide with recent advancements in treating various kinds of Neurological disorders.

This book comprises a new efficient treatment strategy for different kinds of neuronal disorders. It will help in the advancement of alternative treatment scheme for NDDs. In addition, recent nanoparticle-mediated protection for NDDs has also been included in this book. Therefore, the section contains various knowledge that focused on the role of enzyme and polyphenols for PD and AD, respectively.

This book also demonstrates some yoga techniques in the management of NDDs. Moreover, this book explores the natural compound and nanoformulation-based treatment of different NDDs, which are the most advanced treatment options. This book also covers the MS medication strategy by demonstrating the vital effect on animal models. Gut-brain axis based therapy of AD and PD is a hottopic, which is also included in the book chapters. However, Ayurvedic medicine for different NDDs has also been mentioned. Mushrooms mediated treatment of PD and AD is also included for better exploratory knowledge. Thus, we can say that this edition shows distinct advanced treatment alternatives that inevitably attract the interest of scientists and researchers working on NDDs. They can utilize various alternative treatment options for treating neurological disorders. Besides, researchers and scientists all across the world can also use different approaches to the treatment of brain-related ailments disorders. Furthermore, they can also learn separate tools and techniques that have been mentioned in the chapters of the book for NDDs analysis. Thus, it will be a complete package for researchers and scientists working in various fields of NDDs.

Sachchida Nand Rai
Centre of Biotechnology
University of Allahabad
India

ACKNOWLEDGEMENTS

Words are sometimes hard to find when one tries to say thanks for something, as priceless as loving criticism, considerate helpfulness, and valuable guidance. Though facts must be evidently acknowledged, and honest thankfulness must be unequivocally stated. This is what I have humbly attempted to do here.

Above all, I would like to thank the Almighty for making a way and helping me with every step of my life and in the successful completion of this book.

I cannot forget the affection, innumerable blessings, and strength that were bestowed on me by my family. I thank God for giving me wonderful parents Dr. Ravindra Rai and Mrs. Asha Rai, who sincerely raised me with their care and gentle love and have immense faith in me, which brought this work to completion. I would also like to appreciate the support and assistance provided by my uncle and aunty for their consistent love, blessings, and encouragement. I have no words to express my gratitude to my wife Payal Singh, my sister Reena Rai and my brother Ashwini Kumar Rai for their constant support and help during the preparation of this book.

The editor would like to acknowledge UGC Dr. D.S. Kothari Postdoctoral scheme for awarding the fellowship to Dr. Sachchida Nand Rai (Ref. No-F.4-2/2006 (BSR)/BL/19-20/0032).

List of Contributors

Bajaj Priyanka	Institute of Microbial Technology, Sector 39A, Chandigarh- 160036, India
Bajaj Tania	Department of Pharmaceutics, ISF College of Pharmacy, Moga, Punjab-142001, India
Chaturvedi Abhishek Kumar	Central Government Health Scheme, Ministry of AYUSH, New Delhi, India
Chaturvedi Mridula	Amity Institute of Biotechnology, Amity University, Noida, Uttar Pradesh, India
Dewangan Jayant	Genotoxicity lab, CSIR-Central Drug Research Institute, Lucknow, India
Gautam Priyanka	Department of Neurology, Institute of Medical Sciences,Banaras Hindu University, Varanasi, Uttar Pradesh, India
Gupta Nidhi	Department of Psychology, D.D.U. Gorakhpur University, Gorakhpur-273001, Uttar Pradesh, India
Heer Hemraj	Department of Pharmaceutics, ISF College of Pharmacy, Moga, Punjab-142001, India
Jamal Farrukh	Department of Biochemistry, Rammanohar Lohia Avadh University, Faizabad, Uttar Pradesh, India
Jogi Mukesh Kumar	Department of Neurology, Institute of Medical Sciences,Banaras Hindu University, Varanasi, Uttar Pradesh, India
Kaur Vishav Prabhjot	Department of Pharmaceutics, ISF College of Pharmacy, Moga, Punjab-142001, India
Khan Nilofar	Amity Institute of Biotechnology, Amity University, Maharashtra, 410206, India
Kumar Raushan	Department of Biochemistry, University of Allahabad, Allahabad-211002, Uttar Pradesh, India
Kushwaha Ankita	Centre of Biophysics, Ewing Christian College, Prayagraj-211003, India
M.P. Singh	Centre of Biotechnology, University of Allahabad, Prayagraj, India
Pandey Prabhash Kumar	Department of Biochemistry, Faculty of Science, University of Allahabad, Prayagraj, Uttar Pradesh, India
Pathak Abhishek	Department of Neurology, Institute of Medical Sciences, Banaras Hindu University, Varanasi, Uttar Pradesh, India
Patil Ravishankar	Amity Institute of Biotechnology, Amity University, Maharashtra, 410206, India
Rai Sachchida Nand	Centre of Biotechnology, University of Allahabad, Prayagraj, India
Rath Srikanta Kumar	Genotoxicity lab, CSIR-Central Drug Research Institute, Lucknow, India
Sanjay C. Masih	Department of Zoology, Ewing Christian College, Prayagraj-211003, India
Sarma Jayasri Das	Department of Biological Sciences, Indian Institute of Science Education and Research, Kolkata, Mohanpur, Nadia, West Bengal 741246, India

Sengupta Sourodip — Department of Biological Sciences, Indian Institute of Science Education and Research, Kolkata, Mohanpur, Nadia, West Bengal 741246, India

Singh Abhishek Kumar — Amity Institute of Neuropsychology and Neurosciences, Amity University, Noida-201313, Uttar Pradesh, India

Singh Arti — Department of Pharmacology, ISF College of Pharmacy, Moga, Punjab-142001, India

Singh Charan — Department of Pharmaceutics, ISF College of Pharmacy, Moga, Punjab-142001, India

Singh Payal — Department of Zoology, MMV, BHU, Varanasi, India

Singh Ranjan — Department of Biotechnology, Choithram College of Professional Studies, Indore, Madhya Pradesh, India

Tripathi Shambhoo Sharan — Department of Biochemistry, University of Allahabad, Prayagraj-211002, Uttar Pradesh, India

Vivek K. Chaturvedi — Centre of Biotechnology, University of Allahabad, Prayagraj-221002, India

CHAPTER 1

An Introduction to Neurodegenerative Diseases and its Treatment

Payal Singh[1] and **Sachchida Nand Rai**[2,*]

[1] *Department of Zoology, MMV, Banaras Hindu University, Varanasi-221005, India*

[2] *Centre of Biotechnology, University of Allahabad, Prayagraj-221002 , India*

Abstract: In the 21st century, a lot of progress has been made in the treatment against different kinds of Neurodegenerative disorders (NDs). Antioxidant therapy is one of the most common types of therapy for NDs. Among Antioxidant therapy, reduced GSH delivery systems are widely utilized. Gut-microbiome based treatment is also widely accepted. The blood-brain barrier (BBB) is one of the major hurdles that reduce the efficacy of several neuroprotective drugs. That is why nanoformulation based drug is currently trending to potentially treat the neurodegenerative disease. 3D organoid model is employed to mimic the *in vivo* condition for the development of drugs for NDs. Target specific surgical interventions are also utilized to improve the symptoms of neurological diseases. Chemical compound mediated protection only provides symptomatic relief. In long term usage, this chemical compound causes several side effects. Herbal plant-mediated therapy is a better alternative for the same. Diet is a basic part of our life. By manipulating our diet in such a way that include several beans may be very helpful in the treatment of several NDs. Accordingly, this chapter explores some important recent advancement in the treatment of different NDs.

Keywords: Alzheimer's disease, Huntington's disease, *Mucuna pruriens*, Parkinson's disease, Ursolic acid.

INTRODUCTION

In recent years, several targets for different neurodegenerative diseases (NDs) have been identified and tested for therapeutic implications. Different areas of the brain have been explored to find a connection between neuroanatomy and disease progression. Sporadic and genetic level factors have been taken into consideration for therapeutic response against these diseases. Ayurveda provides a very efficient way to prevent progressive degeneration in NDs [1]. The gut-brain axis was explored by several researchers to establish a link between the gut and brain [2]. The following are some advancements made in the treatment of NDs.

* **Corresponding author Sachchida Nand Rai:** Centre of Biotechnology, University of Allahabad, Prayagraj-221002 , India; Tel: +91 9616503505; E-mail: raibiochem@gmail.com

AYURVEDA IN NEURODEGENERATIVE DISEASES (NDS)

Ayurveda plays a very important role in the prevention of different NDs. The progression of several NDs as Parkinson's disease (PD), Alzheimer's disease (AD), Huntington's disease (HD), and Amyotrophic lateral sclerosis (ALS), has been slow down by different Ayurvedic and herbal plant [3]. The bioactive components present in these Ayurvedic and herbal plants are mainly responsible for the underlying therapeutic responses [4]. In PD, *Mucuna pruriens* (Mp) protected the death of dopaminergic neurons in substantia nigra pars compacta (SNpc) and in the striatum (ST) through NF-κB and pAkt1 pathways [5]. The seed extract of Mp contains a significant amount of levodopa (L-DOPA) that provides the major symptomatic response in PD [5, 6]. Ursolic acid (UA) is the major bioactive components in the seed extract of Mp that also shows potent Anti-Parkinsonian activity in the toxin-induced PD mouse model [7, 8]. Similar to Mp, *Withania somnifera* (Ws) also exhibits strong antioxidative activity in the toxin-induced PD mouse model by targeting the apoptotic pathway [9, 10]. Similar to UA, chlorogenic acid (CA) is also found in several herbal plants that exhibit potent anti-oxidative and anti-inflammatory activity in the PD model by modulating the mitochondrial pathways [11, 12]. *Tinospora cordifolia* (Tc) prevented the progressive neurodegeneration in PD by its antioxidative and anti-inflammatory activity in the toxin model of PD [13]. Inflammation is the common characteristics in almost all NDs. Mp inhibits the inflammation in LPS induced in *vitro* cells and might play an important role in the treatment of all NDs [14]. Mp also exhibits its therapeutic activity in the stroke (ischemia) model of rats [15]. The bioactive components of Ws also show therapeutic activity in AD. Withanamides is vital bioactive constituents of Ws that protect from beta-amyloid-induced toxicity in PC12 model of AD [16]. In silico analysis along with integrated system pharmacology shows the potent therapeutic activity of Ws in AD [17]. Ws also shows its therapeutic potential by inhibiting the production of amyloid beta through neuroinflammatory and epigenetic pathways in the AD *in vitro* model [18]. Withanolide is also an important bioactive component in Ws that exhibits neuroprotective activity *via* the intranasal route in the ischemia model of mice [19]. *Gastrodiaelata* (GE) is also an important herbal plant that controls the morphology of mitochondria by attenuating protein aggregations induced by mutant huntingtin [20]. In the 3-nitropropionic acid-induced HD model, seed extracts of *Psoralea corylifolia* Linn. show a neuroprotective effect by improving mitochondrial dysfunction [21]. In the spinocerebellar ataxia 3 cell model, an aqueous extract of *Glycyrrhiza inflata* inhibits aggregation by upregulating PPARGC1A and NFE2L2-ARE pathways [22]. GE also inhibits the aggregation of huntingtin proteins through the activation of the ubiquitin proteasomal system and adenosine A2A receptor [23].

In this way, we can say that Ayurvedic plants and their bioactive components show promising therapeutic activity in different NDs. Further study will be needed to explore the additional Ayurvedic plants and their bioactive components against NDs.

VITAMINS IN NEURODEGENERATIVE DISEASES (NDS)

In this COVID-19 pandemic, vitamins show a very promising response against the viral load [24]. Clinical trials on COVID-19 patients prove that vitamins fight strongly against coronavirus by enhancing host immunity [25, 26]. The neurological symptoms in COVID-19 patients were well managed by vitamin supplementation [27]. Both water-soluble vitamins and lipid-soluble vitamins exhibit immune-enhancing activity and have been tested against different ND, as shown by several types of research. Vitamin D (VitD) improves the cognitive functions in PD, and its low level may be a potential biomarker of mild cognitive impairment [28]. Similarly, ascorbate also improves the cognitive function in PD and decreasesthe urate concentration [29]. Supplementation of Vitamin B9-B12 improves the cognitive functions by neurogenesis in aged rat models who are subjected to gestational and perinatal deficiency of the same vitamins [30]. Vitamins also modulate the progression of AD through multiple pathways [31]. A deficient level of VitD enhances the AD-like pathologies by reducing the antioxidative potential [32]. Vitamin A (VitA) and retinoic acid also improve the cognitive function in cognitive disease [33]. The receptor of retinoic acid is a very important component in all NDs and might be targeted for vitamin supplementation-based therapy [34].

Thus, vitamin supplementation is very vital to improve our immune function and also to manage the neurological symptoms found in different NDs.

GUT-BRAIN AXIS AND ASSOCIATED PRO AND PRE-BIOTICS THERAPY FOR NDS

The dysfunctional gut-brain axis is found in NDs, and it could be an early sign of the disease condition as like in PD, AD, and HD [35 - 37]. Repeated infection of few pathogens like *Citrobacter rodentium is* responsible for the PD pathology in *Pink1$^{-/-}$* mice compared to wild type. Characterization of the gut shows the disturbance in the level of short-chain fatty acids and butyric acid in the PD model *versus* control. Thus, gut-brain homeostasis plays a very important role in PD progression [38]. Probiotics and prebiotics treatment prove to beimproved the homeostasis of different NDs by balancing the activity of the gut-brain axis [39]. The gut microbiome modulates various signaling pathways as it balances the epigenetic pathways in NDs [40]. In diet-induced obese mice, cognitive impairment was significantly alleviated by beta-glucan [41]. Gut dysbiosis is strongly associated with the pathophysiology of HD, and associated pro and pre-biotics therapy considerably improve the disease symptoms [42]. The modulation of various microbiota prevents the progression of AD and offers a significant therapeutic approach to treat this disease [43]. Neuroinflammation was effectively modulated by gut microbiota and prevent progressive neurodegeneration in AD [44]. Neuropathic Pain was influenced by the gut-brain axis by modulating the level of proinflammatory and anti-inflammatory cytokines T cells [45]. Manganese exposure induced neuroinflammation was ameliorated by the gut-microbiota by inhibiting cerebral NLRP3 inflammasome [46]. Progression of MPTP-induced neurodegeneration in the PD model was significantly alleviated by Lactobacillus Plantarum PS128 that restored the normal function of the gut-brain axis [47]. Plant polysaccharides show the ability to modulates the gut-brain axis in different NDs [48]. Lactic acid bacteria improve the deformity in the eye and improves the gut-brain axis in the AD Drosophila model [49]. In salsolinol-induced SH-SY5Y cells, Butyrate protects the progressive neurodegeneration in PD by modulating the gut-brain axis [50].

Therefore, gut-brain axis and associated pro and pre-biotics therapy show strong efficiency in the treatment of different NDs.

NANOPARTICLES AND NANOFORMULATION BASED THERAPY FOR DIFFERENT NDS

The efficacy of different drugs for the NDs shows limited response because of the blood-brain-barrier (BBB) [51]. BBB prevents the delivery of desired drug into the central nervous system (CNS). Strategies like improving the BBB permeability might be dangerous as it allows the delivery of certain undesired molecules also [52]. Nanoparticles and nanoformulation offer an efficient alternative for the hurdle induced by BBB [53]. The size of the nanoparticles is very small, and it easily crosses the BBB. Therefore, the drug is very effective at a very minimum dose as a result of nanoformulation [54 - 56]. Nanotechnology offers a promising response in the treatment of different NDs by providing novel and effective therapeutic approaches for the drug delivery system [57]. In the paraquat (PQ) induced model of drosophila, piperine-coated gold nanoparticles improve the motor response in a significant way [58]. Exosome mediated drug delivery also shows promising advantages over conventional treatment for PD [59]. Intranasal delivery of nanoparticles is more advantageous in different NDs as compared to other routes [60]. Nanoformulation based on herbal drugs also shows potent therapeutic activity in minimum dose in NDs [61]. Curcumin and its nanoformulation exhibit strong therapeutic responses in several neurological disorders [62]. Lipid-based nanoformulation also shows a similar therapeutic activity [63]. In the rotenone-induced PD model, deferoxamine and curcumin loaded nanocarriers prevent the progression of the disease [64]. Anti-amyloid and antioxidant activity were significantly shown by modified magnetic core-shell mesoporous silica nano-formulations with encapsulated quercetin in the AD model [65]. Similarly, in the streptozotocin-induced AD mouse model, the rosiglitazone embedded nanocarrier system offers significant neuroprotection [66]. Likewise, pomegranate and its nano-formulations show a strong therapeutic response in the AD rat model [67].

CONCLUSION

In conclusion, we can say that Ayurveda, herbal plant, and bioactive components show strong therapeutic efficacy with minimal side effects in the treatment of different NDs. Vitamins are very crucial to maintain the normal homeostasis in our body and also enhances the immunity of our body. Both water-soluble and lipid-soluble vitamins are in the hot spot for the treatment of different NDs. The Gut-brain axis and associated pro and pre-biotics therapy also show strong efficiency in the treatment of different NDs. Finally, nanoparticles and nanoformulation based drug delivery show a strong response in the minimum dose that removes the hurdles associated with BBB. These areas are currently hot topics for several researchers working worldwide and offer novel therapeutic targets in the treatment of different NDs. Further studies will be needed to identify other novel approaches and perspectives for the treatment of NDs.

CONSENT FOR PUBLICATION

Not applicable.

CONFLICT OF INTEREST

The authors declare no conflict of interest, financial or otherwise.

ACKNOWLEDGEMENTS

Authors would like to acknowledge UGC Dr. D.S. Kothari Postdoctoral scheme for awarding the fellowship to Dr. Sachchida Nand Rai (Ref. No-F.4-2/2006 (BSR)/BL/19-20/0032).

REFERENCES

[1]   Lakhotia SC. Neurodegeneration disorders need holistic care and treatment - can ayurveda meet the challenge? Ann Neurosci 2013; 20(1): 1-2.
[http://dx.doi.org/10.5214/ans.0972.7531.200101] [PMID: 25205998]

[2]   Carabotti M, Scirocco A, Maselli MA, Severi C. The gut-brain axis: interactions between enteric microbiota, central and enteric nervous systems. Ann Gastroenterol 2015; 28(2): 203-9.
[PMID: 25830558]

[3]   Durães F, Pinto M, Sousa E. Old drugs as new treatments for neurodegenerative diseases. Pharmaceuticals (Basel) 2018; 11(2): 44.
[http://dx.doi.org/10.3390/ph11020044] [PMID: 29751602]

[4]   Xu DP, Li Y, Meng X, *et al.* Natural antioxidants in foods and medicinal plants: extraction, assessment and resources. Int J Mol Sci 2017; 18(1): 96.
[http://dx.doi.org/10.3390/ijms18010096] [PMID: 28067795]

[5]   Rai SN, Birla H, Singh SS, *et al. Mucuna pruriens* protects against MPTP intoxicated neuroinflammation in Parkinson's disease through NF-κB/pAKT signaling pathways. Front Aging Neurosci 2017; 9: 421.
[http://dx.doi.org/10.3389/fnagi.2017.00421] [PMID: 29311905]

[6]   Yadav SK, Prakash J, Chouhan S, Singh SP. *Mucuna pruriens* seed extract reduces oxidative stress in nigrostriatal tissue and improves neurobehavioral activity in paraquat-induced Parkinsonian mouse model. Neurochem Int 2013; 62(8): 1039-47.
[http://dx.doi.org/10.1016/j.neuint.2013.03.015] [PMID: 23562769]

[7]   Rai SN, Yadav SK, Singh D, Singh SP. Ursolic acid attenuates oxidative stress in nigrostriatal tissue and improves neurobehavioral activity in MPTP-induced parkinsonian mouse model. J Chem Neuroanat 2016; 71: 41-9.
[http://dx.doi.org/10.1016/j.jchemneu.2015.12.002] [PMID: 26686287]

[8]   Rai SN, Zahra W, Singh SS, *et al.* Anti-inflammatory activity of ursolic acid in MPTP-induced parkinsonian mouse model. Neurotox Res 2019; 36(3): 452-62.
[http://dx.doi.org/10.1007/s12640-019-00038-6] [PMID: 31016688]

[9]   Prakash J, Chouhan S, Yadav SK, Westfall S, Rai SN, Singh SP. Withania somnifera alleviates parkinsonian phenotypes by inhibiting apoptotic pathways in dopaminergic neurons. Neurochem Res 2014; 39(12): 2527-36.
[http://dx.doi.org/10.1007/s11064-014-1443-7] [PMID: 25403619]

[10]   Prakash J, Yadav SK, Chouhan S, Singh SP. Neuroprotective role of withania somnifera root extract in maneb-paraquat induced mouse model of parkinsonism. Neurochem Res 2013; 38(5): 972-80.
[http://dx.doi.org/10.1007/s11064-013-1005-4] [PMID: 23430469]

[11]   Singh SS, Rai SN, Birla H, *et al.* Neuroprotective effect of chlorogenic acid on mitochondrial dysfunction-mediated apoptotic death of DA neurons in a parkinsonian mouse model. Oxid Med Cell Longev 2020; 2020: 6571484.
[http://dx.doi.org/10.1155/2020/6571484] [PMID: 32566093]

[12]   Singh SS, Rai SN, Birla H, *et al.* Effect of chlorogenic acid supplementation in MPTP-intoxicated mouse. Front Pharmacol 2018; 9: 757.
[http://dx.doi.org/10.3389/fphar.2018.00757] [PMID: 30127737]

[13]   Birla H, Rai SN, Singh SS, *et al.* Tinospora cordifolia Suppresses Neuroinflammation in parkinsonian mouse model. Neuromolecular Med 2019; 21(1): 42-53.
[http://dx.doi.org/10.1007/s12017-018-08521-7] [PMID: 30644041]

[14]   Rachsee A, Chiranthanut N, Kunnaja P, *et al. Mucuna pruriens* (L.) DC. seed extract inhibits lipopolysaccharide-induced inflammatory responses in BV2 microglial cells. J Ethnopharmacol 2021; 267: 113518.
[http://dx.doi.org/10.1016/j.jep.2020.113518] [PMID: 33122120]

[15]   Nayak VS, Kumar N, D'Souza AS, Nayak SS, Cheruku SP, Pai KSR. The effects of *Mucuna pruriens* extract on histopathological and biochemical features in the rat model of ischemia. Neuroreport 2017; 28(18): 1195-201.
[http://dx.doi.org/10.1097/WNR.0000000000000888] [PMID: 28953092]

[16]   Jayaprakasam B, Padmanabhan K, Nair MG. Withanamides in Withania somnifera fruit protect PC-12 cells from beta-amyloid responsible for Alzheimer's disease. Phytother Res 2010; 24(6): 859-63.

[http://dx.doi.org/10.1002/ptr.3033] [PMID: 19957250]

[17]   Hannan MA, Dash R, Haque MN, Choi SM, Moon IS. Integrated system pharmacology and in silico analysis elucidate neuropharmacological actions of Withania somnifera in the treatment of Alzheimer's disease. CNS Neurol Disord Drug Targets 2020; 19(7): 541-56.
[http://dx.doi.org/10.2174/1871527319999200730214807] [PMID: 32748763]

[18]   Atluri VSR, Tiwari S, Rodriguez M, *et al.* Inhibition of Amyloid-Beta Production, Associated Neuroinflammation, and Histone Deacetylase 2-Mediated Epigenetic Modifications Prevent Neuropathology in Alzheimer's Disease *in vitro* Model. Front Aging Neurosci 2020; 11: 342.
[http://dx.doi.org/10.3389/fnagi.2019.00342] [PMID: 32009938]

[19]   Mukherjee S, Kumar G, Patnaik R. Withanolide a penetrates brain *via* intra-nasal administration and exerts neuroprotection in cerebral ischemia reperfusion injury in mice. Xenobiotica 2020; 50(8): 957-66.
[http://dx.doi.org/10.1080/00498254.2019.1709228] [PMID: 31870211]

[20]   Huang NK, Lin CC, Lin YL, *et al.* Morphological control of mitochondria as the novel mechanism of Gastrodia elata in attenuating mutant huntingtin-induced protein aggregations. Phytomedicine 2019; 59: 152756.
[http://dx.doi.org/10.1016/j.phymed.2018.11.016] [PMID: 31004885]

[21]   Im AR, Chae SW, Zhang GJ, Lee MY. Neuroprotective effects of Psoralea corylifolia Linn seed extracts on mitochondrial dysfunction induced by 3-nitropropionic acid. BMC Complement Altern Med 2014; 14: 370.
[http://dx.doi.org/10.1186/1472-6882-14-370] [PMID: 25277760]

[22]   Chen CM, Weng YT, Chen WL, *et al.* Aqueous extract of glycyrrhiza inflata inhibits aggregation by upregulating PPARGC1A and NFE2L2-ARE pathways in cell models of spinocerebellar ataxia 3. Free Radic Biol Med 2014; 71: 339-50.
[http://dx.doi.org/10.1016/j.freeradbiomed.2014.03.023] [PMID: 24675225]

[23]   Huang CL, Yang JM, Wang KC, *et al.* Gastrodia elata prevents huntingtin aggregations through activation of the adenosine A_2A receptor and ubiquitin proteasome system. J Ethnopharmacol 2011; 138(1): 162-8.
[http://dx.doi.org/10.1016/j.jep.2011.08.075] [PMID: 21924340]

[24]   Jayawardena R, Sooriyaarachchi P, Chourdakis M, Jeewandara C, Ranasinghe P. Enhancing immunity in viral infections, with special emphasis on COVID-19: A review. Diabetes Metab Syndr 2020; 14(4): 367-82.
[http://dx.doi.org/10.1016/j.dsx.2020.04.015] [PMID: 32334392]

[25]   Shakoor H, Feehan J, Al Dhaheri AS, *et al.* Immune-boosting role of vitamins D, C, E, zinc, selenium and omega-3 fatty acids: Could they help against COVID-19? Maturitas 2021; 143: 1-9.
[http://dx.doi.org/10.1016/j.maturitas.2020.08.003] [PMID: 33308613]

[26]   Jovic TH, Ali SR, Ibrahim N, *et al.* Could Vitamins Help in the Fight Against COVID-19? Nutrients 2020; 12(9): 2550.
[http://dx.doi.org/10.3390/nu12092550] [PMID: 32842513]

[27]   Shakoor H, Feehan J, Mikkelsen K, *et al.* Be well: A potential role for vitamin B in COVID-19. Maturitas 2020; S0378-5122(20): 30348-0.

[28]   Santangelo G, Raimo S, Erro R, *et al.* Vitamin D as a possible biomarker of mild cognitive impairment in parkinsonians. Aging Ment Health 2020; 1-5.
[http://dx.doi.org/10.1080/13607863.2020.1839860] [PMID: 33111573]

[29]   Spencer ES, Pitcher T, Veron G, *et al.* Positive association of ascorbate and inverse association of urate with cognitive function in people with parkinson's disease. Antioxidants 2020; 9(10): 906.
[http://dx.doi.org/10.3390/antiox9100906] [PMID: 32977491]

[30]   Pourié G, Martin N, Daval JL, *et al.* The stimulation of neurogenesis improves the cognitive status of

aging rats subjected to gestational and perinatal deficiency of B9-12 Vitamins. Int J Mol Sci 2020; 21(21): E8008.
[http://dx.doi.org/10.3390/ijms21218008] [PMID: 33126444]

[31] Alam J. Vitamins: a nutritional intervention to modulate the Alzheimer's disease progression. Nutr Neurosci 2020; 1-18.
[http://dx.doi.org/10.1080/1028415X.2020.1826762] [PMID: 32998670]

[32] Fan YG, Pang ZQ, Wu TY, *et al.* Vitamin D deficiency exacerbates Alzheimer-like pathologies by reducing antioxidant capacity. Free Radic Biol Med 2020; 161: 139-49.
[http://dx.doi.org/10.1016/j.freeradbiomed.2020.10.007] [PMID: 33068737]

[33] Wołoszynowska-Fraser MU, Kouchmeshky A, McCaffery P. Vitamin a and retinoic acid in cognition and cognitive disease. Annu Rev Nutr 2020; 40: 247-72.
[http://dx.doi.org/10.1146/annurev-nutr-122319-034227] [PMID: 32966186]

[34] Clark JN, Whiting A, McCaffery P. Retinoic acid receptor-targeted drugs in neurodegenerative disease. Expert Opin Drug Metab Toxicol 2020; 16(11): 1097-108.
[http://dx.doi.org/10.1080/17425255.2020.1811232] [PMID: 32799572]

[35] Poirier AA, Aubé B, Côté M, Morin N, Di Paolo T, Soulet D. Gastrointestinal dysfunctions in parkinson's disease: symptoms and treatments. Parkinsons Dis 2016; 2016: 6762528.
[http://dx.doi.org/10.1155/2016/6762528] [PMID: 28050310]

[36] Peterson CT. Dysfunction of the microbiota-gut-brain axis in neurodegenerative disease: the promise of therapeutic modulation with prebiotics, medicinal herbs, probiotics, and synbiotics. J Evid Based Integr Med 2020 Jan-Dec; 25: 2515690X20957225.
[http://dx.doi.org/10.1177/2515690X20957225] [PMID: 33092396]

[37] Ghaisas S, Maher J, Kanthasamy A. Gut microbiome in health and disease: Linking the microbiome-gut-brain axis and environmental factors in the pathogenesis of systemic and neurodegenerative diseases. Pharmacol Ther 2016; 158: 52-62.
[http://dx.doi.org/10.1016/j.pharmthera.2015.11.012] [PMID: 26627987]

[38] Cannon T, Sinha A, Trudeau LE, Maurice CF, Gruenheid S. Characterization of the intestinal microbiota during *Citrobacter rodentium* infection in a mouse model of infection-triggered Parkinson's disease. Gut Microbes 2020; 12(1): 1-11.
[http://dx.doi.org/10.1080/19490976.2020.1830694] [PMID: 33064969]

[39] He M, Shi B. Gut microbiota as a potential target of metabolic syndrome: the role of probiotics and prebiotics. Cell Biosci 2017; 7: 54.
[http://dx.doi.org/10.1186/s13578-017-0183-1] [PMID: 29090088]

[40] Kaur H, Singh Y, Singh S, Singh RB. Gut microbiome-mediated epigenetic regulation of brain disorder and application of machine learning for multi-omics data analysis. Genome 2020; 1-17.
[http://dx.doi.org/10.1139/gen-2020-0136] [PMID: 33031715]

[41] Shi H, Yu Y, Lin D, *et al.* β-glucan attenuates cognitive impairment *via* the gut-brain axis in diet-induced obese mice. Microbiome 2020; 8(1): 143.
[http://dx.doi.org/10.1186/s40168-020-00920-y] [PMID: 33008466]

[42] Wasser CI, Mercieca EC, Kong G, *et al.* Gut dysbiosis in Huntington's disease: associations among gut microbiota, cognitive performance and clinical outcomes. Brain Commun 2020 Jul 24; 2(2): fcaa110.
[http://dx.doi.org/10.1093/braincomms/fcaa110] [PMID: 33005892]

[43] Bonfili L, Cecarini V, Gogoi O, *et al.* Microbiota modulation as preventative and therapeutic approach in Alzheimer's disease. FEBS J 2020 Sep 24.
[http://dx.doi.org/10.1111/febs.15571] [PMID: 32969566]

[44] Goyal D, Ali SA, Singh RK. Emerging role of gut microbiota in modulation of neuroinflammation and neurodegeneration with emphasis on Alzheimer's disease. Prog Neuropsychopharmacol Biol Psychiatry 2021; 106: 110112.

[http://dx.doi.org/10.1016/j.pnpbp.2020.110112] [PMID: 32949638]

[45]   Ding W, You Z, Chen Q, *et al.* Gut microbiota influences neuropathic pain through modulating proinflammatory and anti-inflammatory T Cells.. Anesth Analg 2020 Sep 1.
[http://dx.doi.org/10.1213/ANE.0000000000005155] [PMID: 32889847]

[46]   Peterson CT, Yang F, Xin R, *et al.* The gut microbiota attenuate neuroinflammation in manganese exposure by inhibiting cerebral NLRP3 inflammasome. Biomed Pharmacother 2020; 129: 110449.
[http://dx.doi.org/10.1016/j.biopha.2020.110449] [PMID: 32768944]

[47]   Liao JF, Cheng YF, You ST, *et al.* Lactobacillus plantarum PS128 alleviates neurodegenerative progression in 1-methyl-4-phenyl-1,2,3,6-tetrahydropyridine-induced mouse models of Parkinson's disease. Brain Behav Immun 2020; 90: 26-46.
[http://dx.doi.org/10.1016/j.bbi.2020.07.036] [PMID: 32739365]

[48]   Sun Q, Cheng L, Zeng X, Zhang X, Wu Z, Weng P. The modulatory effect of plant polysaccharides on gut flora and the implication for neurodegenerative diseases from the perspective of the microbiota-gut-brain axis. Int J Biol Macromol 2020; 164: 1484-92.
[http://dx.doi.org/10.1016/j.ijbiomac.2020.07.208] [PMID: 32735929]

[49]   Liu G, Tan FH, Lau SA, *et al.* Lactic acid bacteria feeding reversed the malformed eye structures and ameliorated gut microbiota profiles of drosophila melanogaster Alzheimer's Disease model. J Appl Microbiol 2020 Jul 8.
[http://dx.doi.org/10.1111/jam.14773] [PMID: 32640111]

[50]   Getachew B, Csoka AB, Bhatti A, Copeland RL, Tizabi Y. Butyrate protects against salsolinol-induced toxicity in sh-sy5y cells: implication for Parkinson's disease. Neurotox Res 2020; 38(3): 596-602.
[http://dx.doi.org/10.1007/s12640-020-00238-5] [PMID: 32572814]

[51]   Upadhyay RK. Drug delivery systems, CNS protection, and the blood brain barrier. BioMed Res Int 2014; 2014: 869269.
[http://dx.doi.org/10.1155/2014/869269] [PMID: 25136634]

[52]   Dwibhashyam VS, Nagappa AN. Strategies for enhanced drug delivery to the central nervous system. Indian J Pharm Sci 2008; 70(2): 145-53.
[http://dx.doi.org/10.4103/0250-474X.41446] [PMID: 20046703]

[53]   Patel MM, Patel BM. Crossing the blood-brain barrier: recent advances in drug delivery to the brain. CNS Drugs 2017; 31(2): 109-33.
[http://dx.doi.org/10.1007/s40263-016-0405-9] [PMID: 28101766]

[54]   Rizvi SAA, Saleh AM. Applications of nanoparticle systems in drug delivery technology. Saudi Pharm J 2018; 26(1): 64-70.
[http://dx.doi.org/10.1016/j.jsps.2017.10.012] [PMID: 29379334]

[55]   De Jong WH, Borm PJ. Drug delivery and nanoparticles:applications and hazards. Int J Nanomedicine 2008; 3(2): 133-49.
[http://dx.doi.org/10.2147/IJN.S596] [PMID: 18686775]

[56]   Chenthamara D, Subramaniam S, Ramakrishnan SG, *et al.* Therapeutic efficacy of nanoparticles and routes of administration. Biomater Res 2019; 23: 20.
[http://dx.doi.org/10.1186/s40824-019-0166-x] [PMID: 31832232]

[57]   Naqvi S, Panghal A, Flora SJS. Nanotechnology: a promising approach for delivery of neuroprotective drugs. Front Neurosci 2020; 14: 494.
[http://dx.doi.org/10.3389/fnins.2020.00494] [PMID: 32581676]

[58]   Srivastav S, Anand BG, Fatima M, *et al.* Piperine-coated gold nanoparticles alleviate paraquat-induced neurotoxicity in *Drosophila melanogaster.* ACS Chem Neurosci 2020; 11(22): 3772-85.
[http://dx.doi.org/10.1021/acschemneuro.0c00366] [PMID: 33125229]

[59]   Kumar B, Pandey M, Fayaz F, *et al.* Applications of exosomes in targeted drug delivery for the

treatment of Parkinson's disease: a review of recent advances and clinical challenges. Curr Top Med Chem 2020; 20(30): 2777-88.
[http://dx.doi.org/10.2174/1568026620666201019112557] [PMID: 33076810]

[60]   Islam SU, Shehzad A, Ahmed MB, Lee YS. Intranasal delivery of nanoformulations: a potential way of treatment for neurological disorders. Molecules 2020; 25(8): 1929.
[http://dx.doi.org/10.3390/molecules25081929] [PMID: 32326318]

[61]   Moradi SZ, Momtaz S, Bayrami Z, Farzaei MH, Abdollahi M. Nanoformulations of herbal extracts in treatment of neurodegenerative disorders. Front Bioeng Biotechnol 2020; 8: 238.
[http://dx.doi.org/10.3389/fbioe.2020.00238] [PMID: 32318551]

[62]   Mandal M, Jaiswal P, Mishra A. Role of curcumin and its nanoformulations in neurotherapeutics: A comprehensive review. J Biochem Mol Toxicol 2020; 34(6): e22478.
[http://dx.doi.org/10.1002/jbt.22478] [PMID: 32124518]

[63]   Pottoo FH, Sharma S, Javed MN, *et al.* Lipid-based nanoformulations in the treatment of neurological disorders. Drug Metab Rev 2020; 52(1): 185-204.
[http://dx.doi.org/10.1080/03602532.2020.1726942] [PMID: 32116044]

[64]   Mursaleen L, Somavarapu S, Zariwala MG. Deferoxamine and curcumin loaded nanocarriers protect against rotenone-induced neurotoxicity. J Parkinsons Dis 2020; 10(1): 99-111.
[http://dx.doi.org/10.3233/JPD-191754] [PMID: 31868679]

[65]   Halevas E, Mavroidi B, Nday CM, *et al.* Modified magnetic core-shell mesoporous silica nano-formulations with encapsulated quercetin exhibit anti-amyloid and antioxidant activity. J Inorg Biochem 2020; 213: 111271.
[http://dx.doi.org/10.1016/j.jinorgbio.2020.111271] [PMID: 33069945]

[66]   K.C S, Kakoty V, Marathe S, Chitkara D, Taliyan R. Exploring the neuroprotective potential of rosiglitazone embedded nanocarrier system on streptozotocin induced mice model of Alzheimer's disease. Neurotox Res 2020 Jul 18.
[http://dx.doi.org/10.1007/s12640-020-00258-1] [PMID: 32683650]

[67]   Almuhayawi MS, Ramadan WS, Harakeh S, *et al.* The potential role of pomegranate and its nano-formulations on cerebral neurons in aluminum chloride induced Alzheimer rat model. Saudi J Biol Sci 2020; 27(7): 1710-6.
[http://dx.doi.org/10.1016/j.sjbs.2020.04.045] [PMID: 32565686]

CHAPTER 2

Recent Advancement in the Treatment of Neurodegenerative Diseases by Ayurveda

Mridula Chaturvedi and **Abhishek Kumar Chaturvedi***

Amity Institute of Biotechnology, Amity University, Noida, Uttar Pradesh, India and Central Government Health Scheme, Ministry of AYUSH, New Delhi, India

Abstract: Neurodegenerative diseases (NDDs) are not the only diseases but a key term for a range of conditions that mainly affect the neurons in the human brain resulting in progressive degeneration or death of the nerve cells, which is a deadly and debilitating state. It affects millions of people worldwide. The most common NDDs worldwide are Parkinson's disease (PD) and Alzheimer's disease (AD). According to De Lau & Breteler *et al.*, the incidence of PD is about 10 million globally (*i.e.*, approximately 0.3% of the world population) and 1% of those above 60 years. Management of NDDs has become a big challenge in the modern system of medicine & public health at present because of demographic changes worldwide. There is no specific therapy for the conventional management of NDDs in the modern system of medicine. The absence of specific and complete therapy for NDDs in the present era makes Ayurveda more important to consider some alternative and complementary system of medicine for the treatment. Ayurveda is an Indian system of medicine that comes under AYUSH and treats the NDDs since its inception, which is mainly described under the VataVyadhi (neurological disorder) context. In this chapter, the recent advancement in Ayurvedic medicinal plants, RasaAusadhies (herbo-mineral drugs) & combined drugs, *Panchkarma* therapies (bio-purification procedures), and Yoga & Asanas (bodily postures) that successfully treat the various common NDDs worldwide will be described.

Keywords: Alzheimer's diseases, Ayurveda, Herbo-mineral drugs, Neurodegenerative diseases, *Panchkarma*, Parkinson's disease, Yoga.

INTRODUCTION

The building blocks of the nervous system (brain and spinal cord) are neurons that generally do not replace or reproduce if they become dead or damaged, result in

---

* **Corresponding author Abhishek Kumar Chaturvedi:** Central Government Health Scheme, Ministry of AYUSH, New Delhi, India; Tel: +91 8743012029; E-mail: abhishek.bhumedical@gmail.com

Sachchida Nand Rai (Ed.)

problems in the movement known as ataxias or mental function known as dementias [1]. Due to this, they are responsible for the greatest trouble of neurodegenerative disorders in which Parkinson's disease (PD) and Alzheimer's disease (AD) contributes approximately 60-70% of cases worldwide [2]. At cellular, molecular as well as subcellular level, most of the neurodegenerative diseases (NDDs) exhibit the common features [3]. In common NDDs, various intracellular and extracellular changes can be observed, especially in Alzheimer's, Parkinson's, Huntington's, and other NDDs [4]. In the living organism, the cytoplasm and reticulum are mainly conscientious for the fabrication of structural and functional protein molecules for which the mechanism of translational and post-translation synthesis is extremely multifaceted and complicated [5]. The main characteristics of NDDs are amassing of anomalous protein aggregation that leads to inflammation as well as oxidative stress (OS) in the central nervous system (CNS) [6]. These NDDs (PD & AD) are caused by environmental and genetic influences [7].

Scientists recognize that the amalgamation of a person's genes and environment contributes to the threat of developing NDDs. That is, a person may have a gene that makes him more vulnerable to certain NDDs. But how severely the person is exaggerated depends on environmental exposures throughout life [8]. NDDs are exemplified by aggregation of proteins, inflammation, and OS in the CNS, degradation of neurotransmitters in the synaptic cleft due to the elevated activity of enzymes, mitochondrial dysfunction, and excitotoxicity of neurons [9]. Deficiency or inadequate synthesis of neurohormones and transmitters, anomalous ubiquitination, and stress are directly related to NDDs and also some other induced origin including the drugs which are used for the treatment of autism, and other chronic illnesses are not without side effects and injure the blood-brain barrier which leads to various nervous system related disorders [10].

OVERVIEW OF NEURODEGENERATIVE DISEASES (NDDS)

The progressive loss of function as well as the structure of neurons due to known cause or unknown cause, including the death of neuronal cells, are called NDDs. Many NNDs are discovered, which are the result of these degenerative process in which PD, AD, and Huntington's disease (HD) are most common [11]. Such diseases are fatal, not curable, and permanent in nature, resulting in a debilitating situation for the patient. As research works progress, many similarities come into view that linkthese diseases to one another on a sub-cellular level [12].

The Preamble of Common Neurodegenerative Diseases

Several NDDs are discovered since the beginning, but the most common accounts of 70% of cases worldwide are preamble and are discussed below:

Alzheimer's Disease (AD)

The main features of this disease are neuronal inflammation, cognitive decline, neuronal loss, and neuronal death, which are also known as apoptosis. The main etiology of AD is an aggregation of β-amyloid (Aβ). The formation of microtubule associated protein *i.e.* hyper-phosphorylated Tau in the neurons is directly related to the AD [13].

Parkinson's Disease (PD)

This is an example of movement disorder and is characterized mainly by the abnormal accumulation of α-synuclein protein in the neurons [14].

Huntington's Disease (HD)

This disease is a typical NDDs of the CNS and mainly occurs due to the aggregation of abnormal long polyglutamine [15].

NDDs can be generally classified by their scientific presentations, with extrapyramidal and pyramidal movement disorders and cognitive or behavioral disorders being the most frequent. Few patients have pure syndromes, with most having dissimilar clinical features. Although NDDs are classically defined by specific protein accumulations and anatomic susceptibility, they share many elementary processes associated with progressive neuronal dysfunction and fatality, such as proteotoxic stress and its attendant abnormalities in ubiquitin–proteasomal and autophagosomal/lysosomal systems, OS, programmed cell death, and neuroinflammation (Table **1**) [16].

Table 1. Neurodegenerative diseases, clinical features, and etiology & pathological findings.

S. No.	Common Neuro-degenerative Diseases	Clinical Features	Etiology/Pathological Findings	References
1.	Alzheimer's disease (AD)	Commonest NDDs, loss or decrease in memory, alterations in the frame of mind and activities, a most common and frequent cause of dementia, disorientation, and aphasia	Senile or neuritic plaques and neurofibrillary tangles are the main characteristic lesions in affected tissues. Along neuronal axons, Tau protein is normally involved in nutrient transport and directly linked to AD. In AD, the cerebral cortex and hippocampus lobes are severely affected	[17 - 18]

(Table 1) cont.....

S. No.	Common Neuro-degenerative Diseases	Clinical Features	Etiology/Pathological Findings	References
2.	Parkinson's disease (PD)	Second most common NDDs, associated with movement disorders like tremors, rigidity, bradykinesia, and postural instability in both rest position as well as functioning situations	The hammering of dopamine-producing neurons of the mid-brain substantia-nigra Aggregation and deposition of abnormal protein with a deficient clearance of aggregates Impaired mitochondrial function & oxidative stress, inflammation, necrosis, and accelerated apoptosis.	[19]
3.	Huntington's disease (HD)	Atypical NDDs of the CNS are characterized by choreatic movements, dementia, behavioral, affecting, and psychiatric disturbances. Mostly symptoms appear between the ages of 35 - 50 years, although the onset may occur at any time from childhood to old age.	Due to autosomal dominant mutation in an individual's two copies of genes known as huntingtin &aggregation of abnormal long polyglutamine. Expansion of CAG (cytosine-adenine-guanine) triplet repeats in the coding of the gene for the huntingtin protein results in an abnormal protein, which gradually damages cells in the brain	[20]
4.	Amyotrophic lateral sclerosis or motor neuron disease (MND)	It mainly affects the anterior horns of the spinal cord and cerebral cortex. Mainly occurs in 40-60 years of life and rapidly fatal condition within less than 3 years of onset. Neuronal muscle atrophy and all respiratory muscles weakness leads to lung infection, diminished muscle strength and bulk, hyper reflexes, fasciculation, and amyotrophy leads to paralysis and death	Neuronal death due to excitotoxicity which is influenced by glutamate & elevated calcium ions.	[21]

(Table 1) cont.....

S. No.	Common Neuro-degenerative Diseases	Clinical Features	Etiology/Pathological Findings	References
5.	Prion disease	Also known as transmissible spongiform encephalopathy's (TSEs) which is a rare progressive NDDs. Due to long incubation periods, spongiform changes occur with neuronal loss, and malfunction leads to inflammatory response and always fatal	The complete pathology of this disease is not yet understood but it is believed that this disease is due to Prions protein which on abnormal folding leads to brain damage	[22]
6.	Spinocerebellar ataxia (SCA)	Agenetic disorder in which the main symptoms are progressive gait in coordination, along with poor synchronization of hands, speech, and eye movements. Unsteady and maladroit action is due to atrophy of cerebellum and failure of fine association of muscle movements	Pathology is not yet clear but it is a rare inherited neurological disorder ofthe CNS characterized by the slow degeneration of certain areas of the brain.	[23]
7.	Spinal muscular atrophy (SMA)	This is a group of neuromuscular disorders that result in the loss of motor neurons and progressive muscle wasting. Deterioration of muscle bulk, muscle weakness associated with muscle twitching. Deglutition muscles may affect leads to difficulty in swallowing along with scoliosis and joint contractures. Arm, leg, and respiratory muscles are usually affected first.	SMA is due to a genetic defect in the *SMN1* gene in an autosomal recessive manner. The *SMN1* gene encodes SMN, a protein required for the endurance of motor neurons. Loss of these neurons prevents the sending of signals among the brain and skeletal muscles. .	[24]

(Table 1) cont.....

S. No.	Common Neuro-degenerative Diseases	Clinical Features	Etiology/Pathological Findings	References
8.	Multiple sclerosis(MS)	An autoimmune and inflammatory demyelinating disease of the CNS in which the insulating covers of nerve cells in the brain and spinal cord are damaged. Due to this damage the capability of parts of the nervous system to transmit signals disrupted, resulting in a series of symptoms, including physical, mental, and psychiatric problems. Specific symptoms can include double vision, blindness in one eye, muscle weakness, and trouble with sensation or coordination.	Yet the cause is not clear, the underlying mechanism is thought to be either destruction by the immune system or failure of the myelin-producing cells.	[25]

Table 2. Five mechanism of pathogenesis of neurodegenerative diseases.

Mechanism of the Pathogenesis of Neurodegenerative Diseases				
Deficiency of Proteins	Dysfunction of Proteins	Aggregation of Proteins	Mutation of Genes	Misfolding of Proteins
Deficiency of ubiquitin–proteosome–autophagy system, OS, free radical formation, mitochondrial dysfunction, and impaired bioenergetics is the main cause of NDDs.	Neuronal malfunction and inflammation lead to disruption of Golgi apparatus and	The abnormal interface of intracellular and extracellular self aggregating misfolded	Formation of several gene-gene complexes and gene-environmental interaction with a mutation in the genes encoding protein constituents	Inappropriate folding or misfolding protein Anatomical and functional changes of a normal protein Formation of protein aggregates through different

(Table 2) cont.....

	axonal transport which is due to dysfunction of neurotrophins.	proteins deposition along with the formation of high ordered insoluble fibrils leads to a variety of pathological characteristics involve in NDDs.	resulted in dysfunction and death of neuronal as well as glial cells.	supra-molecular organizations (O-2) during the electron transport chain Manganese superoxide dismutase Hydrogen peroxide (H_2O_2) in the mitochondria glutathione peroxidase H_2O Increase in reactive oxygen species(ROS) resulting in a decrease in the mitochondrial membrane potential, loss of ATP, energy collapse, and subsequent cell death

Pathogenesis of Neurodegenerative Diseases

The main features of NDDs are progressive dysfunction of the specific characteristic of neurons and neuronal defeats which are associated with extra and intracellular accumulation of misfolded proteins. There are mainly five mechanisms of pathogenesis that lead to NDDs (Table **2**) [26 - 27].

Pathway of Development of Neurodegenerative Diseases

The pathway involved in the generation of NDDs is shown in the ray diagram below in Fig. (**1**) [28].

Common Signs & Symptoms of Neurodegenerative Diseases

These signs & symptoms are common in almost all NDDs in which some are specific, and some are nonspecific [29 - 30]. Fig. (**2**) shows the common sign and symptoms of NDDs.

Diagnosis of Neurodegenerative Diseases

A number of tests are now emerged and commonly used in the medical practice for the detection of various brain disorders, which helps to find out the correct diagnosis for the NDDsand their correct management (Table 3) [31].

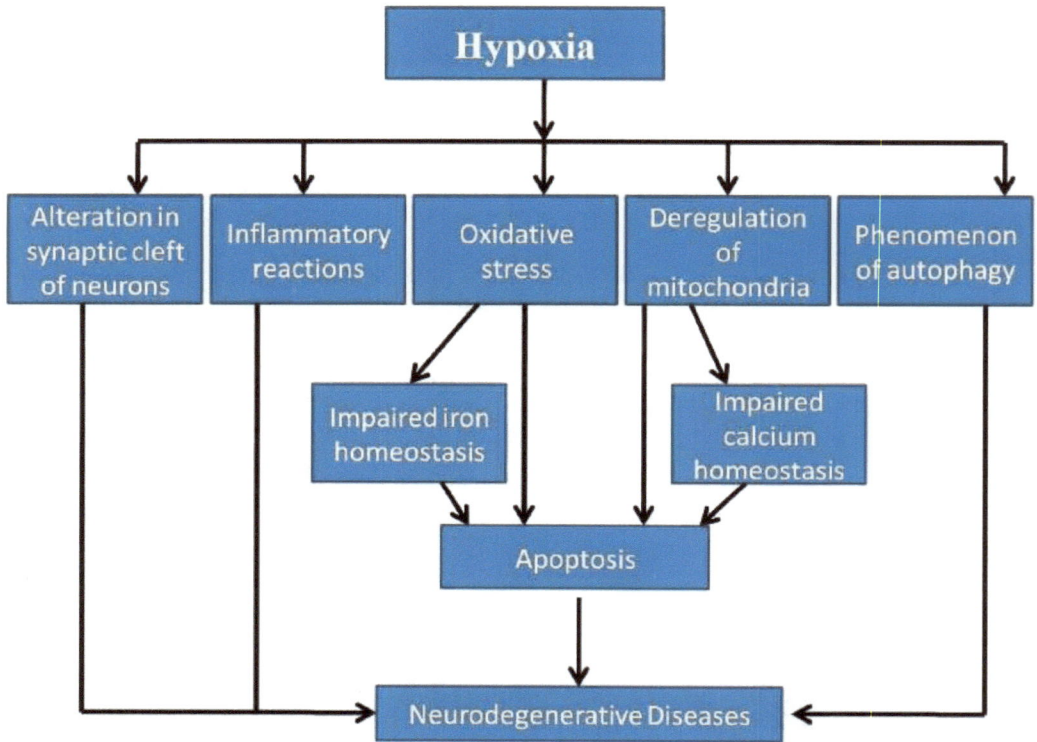

Fig. (1). Pathway of the genesis of NDDs.

Table 3. Diagnostic tests of neurodegenerative diseases.

S. No.	Purpose of Test	Name of Diagnostic Test
1.	For the screening of NDDs	• Addenbrooke's cognitive examination • Frontal assessment battery
2.	For the testing of the function of languages	• Boston or Graded naming test • Pyramids and palm trees test • Trog test • Peabody vocabulary test • Token test

(Table 3) cont.....

S. No.	Purpose of Test	Name of Diagnostic Test
3.	For the testing of memory	• Wechsler memory scale • Adult memory and information processing battery • Rivermead behavioral memory test • List learning tests • Recognition memory test • Doors and people test • Autobiographical memory interview
4.	For the testing of functions of pre-morbidity	• National adult reading test • Wechsler test of adult reading
5.	For the testing of visuoperceptual functions	• Visual and object space perception battery • Benton line orientation test • Birmingham object recognition battery • Behavioral inattention test
6.	For the testing of intelligence	• Wechsler adult intelligence scale
7.	For the testing of executive functions and attention	• Behavioral assessment of dysexecutive syndrome • Deliskaplan executive function system • Wisconsin card sorting test • Trail making test • Stroop test • Hayling and Brixton tests • Verbal fluency

Common sign and symptoms of Neurodegenerative disorders

1. Loss of memory
2. Muscular rigidity
3. Sleep disorders
4. Depression
5. Postural & pace impairment
6. Psychiatric symptoms
7. Constipation

8. Bradykidnesia
9. Cognitive impairment
10. Pain & fatigue
11. Rest tremor
12. Autonomic dysfunction
13. Olfactory dysfunction
14. Mood disorders

Fig. (2). Sign & Symptoms of NDDs.

A RANGE OF TREATMENT FOR THE NEURODEGENERATIVE DISEASES

Gradual and progressive neuronal loss in the brain is the main characteristic of

NDDs which leads to cognitive impairments, memory loss, motor function deficiency, and even death. In developed as well as in developing countries, due to the increasing rise in the case of NDDs, the longevity of the population is affected. The prevalence rate of AD in 2006 worldwide was 26 million, and by 2050 the prevalence rate will become 4 times. However, the prevalence rate of PD becomes 2 times at the end of the year 2030 [32]. Due to this increased rate of NDDs worldwide, the development of effective and definite therapeutic strategies is one of the essential scientific challenges.

The Modern System of Medicine

The western system of medicine is a miracle and quick effect system of medicine, which includes surgical and medical management, both but not yet developed to completely cure the neurodegenerative disease.

Medical Management

These NDDs are illness which has very high morbidity as well as mortality rates, especially PD, AD, and HD. The main features of most of the NDDs are progressive neuronal cell death and diminished muscle bulk and weakness. Several types of research had been done to understand the process of this neuronal cell death, but due to own mechanism of neuronal cell death, the sign and symptoms of these NDDs are different from the specific existed pathway of cell death [33]. Due to these unique features, the management of NDDs leads to novel therapeutic strategies. The conservative therapies for NDDs, for example, L-dopa for PD and cholinesterase inhibitors for AD, are directed at treating only the neurological symptoms but have no or very mild effects on disease progression (Table 4) [34]. There is only symptomatic treatment available in the modern system of medicine for the NDDs.

Table 4. Medicaul management for nerodegenerative diseases.

S.No.	Sign & Symptoms/Neurodegenerative Disease	The Modern System of Medicine	References
1.	Parkinson disease & movement disorders	Dopaminergic drugs	[35]
2.	Cognitive disorders	Cholinesterase inhibitors	[35]
3.	Behavioral and psychological symptoms of dementia	Antipsychotic drugs	[35]
4.	Pain & fatigue	Analgesic drugs	[35]
5.	Associated infections	Anti-inflammatory drugs & antibiotics	[35]
6.	Tremor and refractory movement disorders	Deep brain stimulatory drugs	[35]
7.	Cerebellar ataxia & Huntington's disease	Riluzole	[35]

(Table 4) cont.....

S.No.	Sign & Symptoms/Neurodegenerative Disease	The Modern System of Medicine	References
8.	Alzheimer's disease	NSAIDs [Non-steroidal anti-inflammatory drugs)	[35]
9.	Neuroprotection in Parkinson's disease	Caffeine A2A receptor antagonists & CERE-120 (Adeno-associated virus serotype 2-neurturin)	[35]

Surgical Management

A jumble of surgical methods like neural transplantation, gene therapy, neurosurgical therapy, deep brain stimulation, *etc.*, are discovered, but they are also limited and not proven to effective so far [36].

Emerging Management

A variety of other therapies are also discovered for the management of NDDs, like stem cell transplantation, immunotherapy, and nanotherapy but still not prove yet [37].

Ayush or Indian System of Medicine

Nowadays, there is no specific complete treatment in the modern system of medicine that can cure degenerative diseases. However, the western system of medicine is not highly developed to treat NDDs completely and remains too many problems arise to handle the serious and progressive symptoms of these diseases. The chemical or synthetic drugs are very expensive and acts as only symptomatic and extensive treatment, which shows severe and inevitable side effects on the human brain with poor patient compliance [38]. Due to the absence of complete cure and specific treatment of these NDDs, there is a need for alternate and complementary systems of medicine for treatment, *i.e.*, traditional medicine, Indian medicine, or AYUSH system of medicine. The herbal and AYUSH system of medicines are favored over human-made drugs for a range of human brain disorders including, AD, PD, depression, anxiety, *etc.* The Ayurveda system of Indian medicine is well accepted by health care professionals as well as people worldwide, which played a significant role since 5000 years ago to manage the prominent & elevated factors and by improving the lifestyle for good survival of these patients. Traditionally in several neurological conditions, the Ayurvedic system of medicine has been used because of its economic, nil side effects, and accessibility, which offer considerable advantages. This generates positive hopes for NDDs patients towards the conventional established systems of herbal remedies. It is expected that more than 60 million Indian populations suffer from mental disorders, while the country lags far behind the world for treatments and costs in the hospitals for mental cure [39]. Ayurveda does not only treat the acute symptoms of these diseases, but it also considers all of the related signs& symptoms and tries to get better the quality of activity of daily life for the patients. Ayurvedic medicine can sometimes exhibit protective effects or slow the morbidity of these diseases. Ayurvedic medications have been specifically used for the management of disorders and regain neuronal function and found to significantly relieve the sign and symptoms of NDDs [40]. The Ayurvedic management includes the treatment range for single plant drugs or combined drugs as well as Yoga and biopurification therapies. The parts of the plant used for the treatment of these NDDs varies from diverse plant parts, ranging from the whole plant, roots, stem, bark, leaves, flowers, fruits to seeds while the chemical compounds from these plant source varied from straight-chain fatty acids to terpenoids, steroids, flavonoids, alkaloids, peptides, *etc.*, [41].

RECENT ADVANCEMENT IN THE AYURVEDIC MANAGEMENT OF NEURODEGENERATIVE DISEASES

Fig. (**3**) shows three doshas of the human body as per Ayurveda. In Ayurvedic textbooks, NDDs are mentioned as *Vata-Vyadhi* influenced broadly by the unbalanced consequence of *"Vata"*humour, identified as one most important constituent of *"Tridosha"*; a determinant of a person's basic constitution type called *"Prakriti"*. According to Ayurveda, there are three manifestations of the *'Vata'* diseases, namely *'VataVriddhi'* (neural hyperfunctioning), *'VataKsaya'* (neural hypofunction), and *'Avarana'* (masked functioning). In all Ayurvedic test book, it is mentioned that the majority of *Vata* disease is incurable, but *Acharyas* (Ayurvedic masters) have described the diverse management & treatment for these NDDsvery clearly, which makes life easier and increases the life expectancy of these patients, which includes *'Panchkarma'* (five actions of biopurification), Yoga, herbal medicine &herbo-mineral medicine. The management aim for the cure of NDDs is to first alleviate the symptoms, check complications, and stoppage neurons destruction [42]. Fig. (**4**) shows the management of NDDs by Ayurveda.

Fig. (3). Three Doshas of the human body according to Ayurveda.

Ayurvedic/herbal Medicinal Plants

Currently, a novelistic medication system Ayurveda is considered as growing therapy for the complete anticipation of neurodegeneration. This medicine system consists of herbal/Ayurvedic medicinal plants, which are traditionally used as folker medicine that developed over generations within various societies before the era of modern medicine. These Ayurveda gems or medicinal plants are known since 5000 years ago and used in practices of traditional practice very frequently in India and other parts of the world for the treatment of various ailments. These medicinal plants synthesize hundreds of several chemical and active compounds & phytochemical, which show several biological activities and pharmacological action against various diseases [43]. These Ayurvedic medicinal plants consist of active compound constituents who show neuroprotective properties by inhibition of cholinesterase activity, antioxidative properties, and anti-inflammatory activities because active acetylcholine, which discharges in the brain region, is the main part for the treatment and cure of NDDs. The pathological condition of CNS is characterized by neurofibrillary tangles, depletion of neurotransmitters in the neurons and synaptic cleft are majorly liable for the breakdown of acetylcholine in the synaptic region, and low levels of acetylcholine is related to age-related disorders that lead to loss of cognitive ability [44]. Due to OS, ROS developed in the genetic system can contribute to the break of biological macromolecules, and as a result, pathological conditions at the cellular level can become more apparent [45].

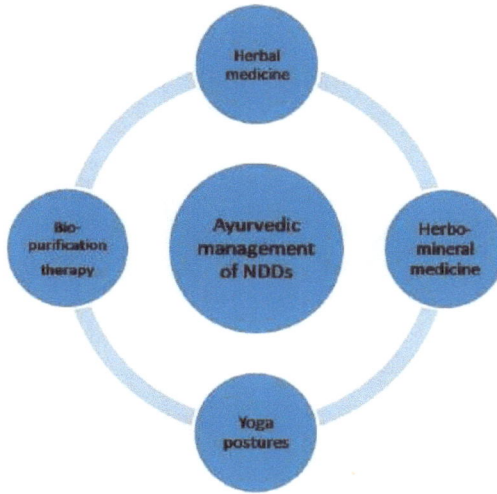

Fig. (4). Ayurvedic management of NDDs.

The research in the field of Ayurvedic medicinal plants has industrialized into the investigation of accepted active compounds responsible for antioxidative and anti-aging properties that can also be useful for NDDs [48]. It is important to excite the cholinergic receptors in the CNS or increase the prolonged production of acetylcholine in the synaptic cleft with the help of such dynamic constituents that could slow down the activities of AChE and BChE in the neuronalsystem [49]. Figs. (**5 - 13**) is a different Ayurvedic plant that is used to manage neurodegenerative diseases collected from Banaras Hindu University, Varanasi.

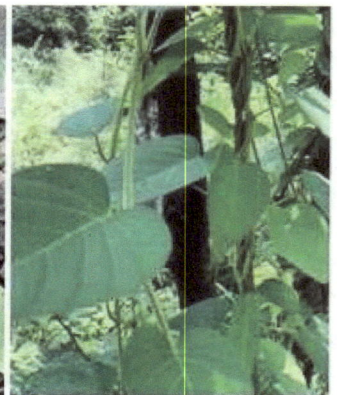

Fig No. 5   Fig No. 6   Fig No. 7

(Figs. 5-13) contd.....

Fig No. 8   **Fig No. 9**   **Fig No.10**

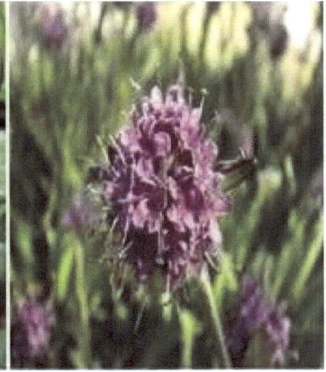

Fig No.11   **Fig No. 12**   **Fig No. 13**

Figs. (5-13). Management of Neurodegenerative diseases by using different Ayurvedic plants.

Table 5. Selected Ayurvedic medicinal plants for the treatment of NDDs.

S. No.	Botanical Name	Common Name	Family	Parts Used	Doses	Fig. No.	References
1.	*Withania somnifera*	*Ashwagandha*	Solanaceae	Root	500 mg/kg	5	[46]
2.	*Centella asiatica*	*Mandukparni*	Apiaceae	Whole plant	400 mg/kg	6	[46]
3.	*Tinospora cordifolia*	*Guduchi*	Menispermaceae	Stem	350 mg/kg	7	[46]
4.	*Convolvulus pluricaulis*	*Shankhpushpi*	Convolvulaceae	Root	300 mg/kg	8	[46]
5.	*Celastrus paniculatus*	*Jyotishmati*	Celastraceae	Root	400 mg/kg	9	[47]

(Table 5) cont.....

S. No.	Botanical Name	Common Name	Family	Parts Used	Doses	Fig. No.	References
6.	*Mucuna pruriens*	*Atmagupta*	Fabaceae	Seeds	350 mg/kg	10	[47]
7.	*Curcuma longa*	*Haridra*	Zingiberaceae	Root	100 mg/kg	11	[47]
8.	*Bacopa monnieri*	*Brahmi*	Scrophulariaceae	Root	300 mg/kg	12	[47]
9.	*Nardostachysjatamansi*	*Jatamansi*	Caprifoliaceae	Root	400 mg/kg	13	[47]

Source: Herbal Garden, Faculty of Ayurveda, Banaras Hindu University, Varanasi, UP, India.

Table 6. Properties of selected Ayurvedic medicinal plants for the treatment of NDDs.

S. No.	Medicinal Plant	Main Active Compounds	Neuroprotective Properties	Specific Disorder	References
1.	*Withania somnifera*	WithaferinASitoindosideIX PhysagulinD WithanosideIV Viscosalactone	Antioxidative Anticholinesterase	Schizophrenia Movement disorders	[50]
2.	*Centella asiatica*	Asiaticoside	Anticholinesterase	Schizophrenia Epilepsy	[51]
3.	*Tinospora cordifolia*	Berberine Choline Tembetarine Tinosporin Palmitine Jatrorrhizine	Anti-Glutamate induced excitotoxicity	Ischemia Epilepsy Parkinson disease	[52]
4.	*Convolvulus pluricaulis*	Cinnamic acid Silane Tropane alkaloids	Antioxidative Ameliorating Apoptosis	Parkinson disease	[53]
5.	*Celastrus paniculatus*	Pristimerin β-amyrin β-sitosterol	Antioxidative	Depression Cognitive impairment	[54]
6.	*Mucuna pruriens*	Levodopa Tetrahydroisoquinoline	Anti-inflammatory Neuroprotective Anticholinesterase	Parkinson disease	[55]
7.	*Curcuma longa*	Curcumin	Antioxidative Anti-inflammatory	Alzheimer's Parkinson's Stroke	[56]
8.	*Bacopa monnieri*	Bacoside Brahmin Herpestine	Anticholinesterase	Memory loss Alzheimer disease	[57]

(Table 6) cont.....

S. No.	Medicinal Plant	Main Active Compounds	Neuroprotective Properties	Specific Disorder	References
9.	*Nardostachysjatamansi*	Angelicin β-eudesmol Nardol	Antioxidative Anticholinesterase	Parkinson disease Schizophrenia	[58]

Different Actions of an Extract Derived from Ayurvedic Medicinal Plants

Recently, medicinal plants have gained wide recognition because of their fewer side effects compared to man-made medicines and necessary to meet the requirement of medicine for increasing the human population. However, a steady supply of source material often becomes difficult due to various factors like diverse geographical distribution, environmental changes, cultural practices, labor cost, and an assortment of superior plant stock, and over-exploitation by pharmaceutical industries (Tables **5** and **6**) [59].

Withania somnifera: This medicinal plant is one of the most important plants of the Ayurvedic system of medicine and useful in the treatment of various kind of disease and specially used in neurological disorders, which contains a range of chemical compounds mainly with A-ferin, withanone, and other flavonoids which show evidence of strong anti-oxidant properties, synergistic effect and having the efficiency to increase catecholamines level and regulation of apoptotic processes [60]. In the substantia nigra pars compacta region of the mid-brain in Parkinsonian mice models, it also shows neuroprotection of dopaminergic neurons [61]. Due to the ability to ameliorate the level of BPA intoxicated oxidative stress, thereby potentially treating cognitive dysfunction, which acts as the major admonition sign in manyNDDs [62]. Another study reveals that Methanol: Chloroform (3:1) extract made from the dried roots of W. somnifera and subjected to LC-MS analysis showed a neuroprotective effect against β-amyloid-induced cytotoxicity and HIV-1 infection [63].

Centella asciatica: A well known Ayurvedic medicine and is indicated for a variety of illness which have budding neuroprotective properties. The recent studies confirmed that *in vitro, C. Asiatica* acts as an antioxidant and diminish the effect of oxidative *stress* and promotes dendrite arborisation and elongation, and also protects the neurons from apoptosis, while *in vivo* studies it is shown that the whole extract and individual compounds have a protective effect against various neurological diseases mainly on AD, PD, learning and memory enhancement, neurotoxicity and other mental illnesses such as depression, anxiety and epilepsy [64]. However, the capability of *C. asiatica* in enhancing neuroregeneration has not been studied much and is limited to the rejuvenation of crushed sciatic nerves and safety from neuronal injury in hypoxia conditions [65]. On the study of leaf extract of *C.asciatca* for expressing normal human alpha-synuclein (h-αS) in the neurons on the climbing ability, activity pattern, lipid peroxidation, protein carbonyl content, glutathione content, and glutathione-S- transferase activity in the brains of transgenic Drosophila model flies which showed a significant delay in the loss of climbing ability and activity pattern and reduced the oxidative stress in the brains of Parkinson's drosophila (PD) flies as compared to untreated PD flies which suggested that its leaf extract is potent in reducing the PD symptoms in transgenic Drosophila model of Parkinson's disease [66].

***Tinospora cordifolia*:** This is one of the miracle drugs in the Ayurvedic system of medicine that possess various medicinal properties to treat various ailments, including fever, inflammation, pain, asthma, and epilepsy [67]. In Ayurveda, *Tinospora cordifolia* is known as a learning and memory enhancer. It enhances it by immune-stimulation and increasing acetylcholine synthesis [68]. Aqueous extract of *this plant* roots showed verbal learning and logical memory enhancement [69]. Various properties of *T. cordifolia* may also be involved in neuroprotection, such as the binding and detoxification of metal ions or free radical scavenging [70].

***Convolvulus pluricaulis*:** A very common memory enhancer drug and is used in frequent practice of the Indian system of medicine. Research has shown that the aqueous extract and ethyl acetate of this plant increases memory functions and learning abilities [71]. By regulating the stress hormone production in the body, it calms the nerves [72]. Organization of *Convolvulus pluricaulis* increased acetyl-cholinesterase activity in hippocampal CA1 and CA3 regions associated with memory utility and learning abilities [73]. Ethanolic extract of this plant also significantly enhanced learning abilities and memory retention in rats [74].

***Celastrus paniculatus*:** *Jyothismati* oil from seeds of *Celastruspaniculatus*, is extensively used in the native medicinal systems to treat brain-related disorders. A study reveals that on the administration of seed oil (400 mg/kg) to the Wistar rats, a decrease in AChE activity was noted in the treated animals leading to increased cholinergic activity in the brain.To study the effect on learning and memory, a radial arm maze paradigm was used. No side effects were observed with the administration of the seed oil [75].

***Mucuna pruriens*:** Unadventurously, *Mucuna pruriens* seeds are used for maintaining male fertility in India, but it is used as a rejuvenator drug having neuroprotective property. To date, modern drugs that have been used to manage PD have only shown symptomatic relief with several unpleasant effects besides their inability to prevent neurodegeneration [76]. Neuroinflammation plays an important role in the advancement of PD and can be targeted for its effective treatment. Study findings suggested that effects of mice intoxicated with 1-methyl-4-phenyl-1,2,3,6-tetrahydropyridine (MPTP) have a significant increase in inflammatory parameters like glial fibrillary acidic protein, inducible nitric oxide synthase, intercellular cell adhesion molecule, and tumor necrosis factor-alpha in substantia nigra pars compacta of parkinsonian mice while aqueous extract of Mucuna (100 mg/kg) orally administered to these rats has notably reduced these inflammatory parameters. It also inhibited the MPTP induced activation of NF-κB and promoted pAkt1 activity, which further prevented the apoptosis of the dopaminergic neurons and exhibited significant antioxidant defense by inhibiting the lipid peroxidation and nitrite level and by improving catalase activity and enhancing GSH level in the nigrostriatal region of mouse brain [77]. The therapeutically main constituent of ailment for tremor disease is the levodopa content of *Mucuna pruriens*. The endogenous accretion of L-dopa, on a dry weight basis, in tissue-cultured *M. pruriens* plant cells was found to be in a range of 0.2 to 2.0% [78].

Curcuma longa: A most common spice in the Indian kitchen as well as a common Ayurvedic herb, possesses anti-inflammatory activity and reduces Alzheimer's risk.It reduces the deposition of plaque in the brain and decreases oxidative stress and amyloid pathology [79]. Epidemiologic studies showed that in Asian countries where turmeric is frequently utilized in diet, there are 4.4 fold lower cases of Alzheimer's [80]. A study reported that low doses of Curcumin reduced Aβ level up to 40% in mice with AD as compared to control drugs and caused a 43% decrease in plaque burden that this Aβ has on the brain of mice with AD [81].

Bacopa monnieri: It is commonly utilized in Ayurvedic medicine and acts as a nerve tonic, cardiotonic, diuretic, and as a therapeutic agent against asthma, insomnia, epilepsy, and rheumatism [82]. This plant has antioxidant properties & utilized for reminiscence and cognitive function enhancement [83]. Extracts of *B. monnieri* has been extensively investigated for their neuropharmacological effects and nootropic actions [84]. In the hippocampus, *B. monnieri* increases the protein kinase activity and inhibited cholinergic degeneration, and shows improved cognition effect in Alzheimer's model of rat [85].

Nardostachys jatamansi: It contains sesquiterpene valeranone that has been used for the treatment of stress [86]. In a study, *Nardostachysjatamansi* exhibited memory maintenance and learning enhancing abilities in aged and young mice and reversed scopolamine and diazepam induced amnesia as well as also reversed aging-induced amnesia *and helps* in the prevention of stress-induced memory deficit [87].

However, several kinds of research are also in process on other Ayurvedic medicinal plants, which shows neuroprotection activity by constituting different chemical compounds and having different properties is under investigation like *Emblica officinalis, Adhatodavasica, Piper nigrum, Tribulus terrestris, Terminalia arjuna, Ginkgo biloba, etc.,* [88].

Herbo-mineral Drugs

Various research studies were conducted on herbo-mineral drugs to show their neuroprotective effect on NDDs. One research study was conducted in the cytogenetics laboratory in Banaras Hindu University on *Ayurvedic AmalakiRasayana* (A preparation derived from Indian gooseberry fruits) and *Rasa-Sindoor* (An organo-metallic *Bhasma* prepared from mercury and sulfur), which suppress neurodegeneration in fly models of HD and AD. This study finding reveals that these Ayurvedic formulations facilitate 'healthy aging', for their function in neuroprotection in fly models of polyQ (127Q and Huntington's) and AD. The nutritional supplement of these formulations during the larval period substantially suppressed neurodegeneration in fly models of polyQ and AD without any side-effects. Dietary *AmalakiRasayana* or *Rasa-Sindoor* prevented accumulation of inclusion bodies and heat shock proteins, suppressed apoptosis, increased the levels of heterogeneous nuclear ribonucleoproteins, and cAMP response element-binding protein and at the same time enhanced the ubiquitin–proteasomal system for better protein clearance in affected cells [89]. A study was conducted on *Swarnprashana* (A preparation of gold *Bhasma)* on the hippocampal lobes of 18 Albino rats by administering the preparation of gold with the control group and after study findings suggested that the control group has normal cellularity in the brain while rats with drugs have increased cellularity in dentate gyrus [90]. *Brahmi Ghrita* was claimed for the treatment of learning and memory disorders in human beings. On assessing learning and memory activity of this drug with the control group and Piracetam on rats, findings suggested that on the evaluation of learning and memory, the activity of *Brahmi Ghrita* using elevated maze plus and passive avoidance test, piracetam treated rats demonstrated a significant decrease in transfer latency in modified elevated plus maze test and increase in step-through latency in passive avoidance test compared with control rats in a dose-dependent manner [91]. According to the Ayurvedic text book, some Ayurvedic *Bhasmas* and drugs like *Rajat Bhasma* (Silver nanoparticles), which have *Vata-Shamak, Madhura Vipaka, Kashaya-Amla Rasa, Sheetala, Snigdha properties,* and *also Brimhana,* so it plays a significant role in the nervous system, and *Smiritisagar Rasa* has *Tikshna, Ushna, Vyavayi* as well as *Yogvahi* properties, *Kapha-vataShamak* and *Bhawna Dravya* are *Brahmi, Vacha, Malkangni* also supportive in mental slowness and misery [92]. *Saraswatarishta* is a herbo-mineral preparation, which is *Medhyarasayanas.*

Evaluating the antidepressant effect of *Saraswatarishta*alone and in combination with imipramine and fluoxetine in animal models of depression its finds that individual drugs and combinations produced a significant decrease in immobility time. However, values for the combination of fluoxetine with were found to be lesser than that for individual agents, and combination of *Saraswatarista*with imipramine did not enhance its anti-depressant effect in any of the parts [93].

Various poly hero-mineral drugs are commonly used in the Ayurveda practices for the management of dementia, and other NDDsand showed significant improvement to pacify the symptoms (Table **7**) [94]. Fig. (**14**) shows diverse and different herbal and mineral preparations in Ayurveda.

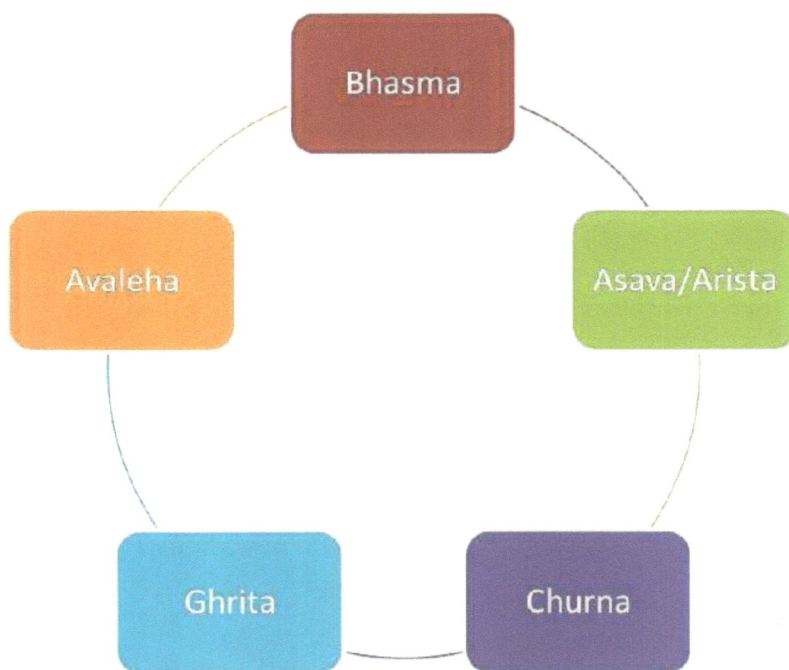

Fig. (14). Different forms of herbo-mineral preparations in Ayurveda.

Table 7. Different forms of herbo-mineral drugs used in the treatment of NDDs.

S.No.	Type of Formulation	Name of Formulation	Traditional Usage
1.	*Bhasma* (Metallic calx preparations)	*Mukta Pisti* *Swarna Bhasma* *Rajat Bhasma* *Smiriti Sagar Rasa* *Kumar Kalyan Rasa*	*Piitaj* mental disorders as well as various forms of dementia

(Table 7) cont.....

S.No.	Type of Formulation	Name of Formulation	Traditional Usage
2.	*Ghrita* (Medicated cow Ghee preparations)	*Brahmi Ghrita* *KushmandakaGhrita* *Siddharthakghrita*	*MedhyaRasayana* (nootropic) insanity, mental retardation, and dementia
3.	*Asava /Arista* (Fermentation products)	*Sarasvatarista* *Ashwagandharista*	Mental weakness and other various mental disorders
4.	*Churna* (Medicated herbal powder)	*Saraswat Churna* *Brahmi Churna* *ShankhpuspiChurna*	Promote intellect, memory, and awareness
5.	*Avaleha* (Medicated electuaries)	*Brahmairasayna* *Jyotismatirasayna* *Swarna PrashanLeh*	Mental debilities

Panchkarma (Bio-purification Procedures)

In Ayurveda, *VataVyadhi* (neurological disorders) are considered as chief serious diseases because of their severity and cured by rigorous *Panchkarma* and palliative medicines by a skilled physician. NDDs are mentioned in Ayurveda in the same context in which PD is named *KampaVata* due to the same clinical features, as mentioned in the modern system of medicine [95]. Due to the genetic and environmental causes, there is an increased incidence of NDDsthroughout the world [96]. *Panchkarma* is an effective procedure in treating a wide range of NDDs. Comprehensive Ayurveda system of medicine by its holistic approach of *Panchkarma*, palliative medicines, diseases specific *Rasayana* (Immuno-modulators), dietary recommendations, and lifestyle recommendations can provide definitive and long-lasting results [97]. *Panchkarma* therapy not only eliminates the disease-causing toxins but also revitalizes the tissues and has a full remedy role as a promotive, protective & restorative modality [98].

Overview of Panchkarma Therapy

It is a fivefold treatment modality which is divided into three parts, namely *Poorva Karma* (preparatory regimen), *Pradhan Karma* (main operative regimen), and Paschat Karma (post-operative regimen). *Panchkarma* procedure purifies various systems of the human body and expels out the accumulated toxic metabolites from the body [99]. Basically, *Panchkarma* is a bio-cleansing procedure that detoxifies the body and helps in increasing the bio-availability of drugs, diet, *etc.*, [100]. Fig. (**15**) shows three vital procedures of *Panchkarma*.

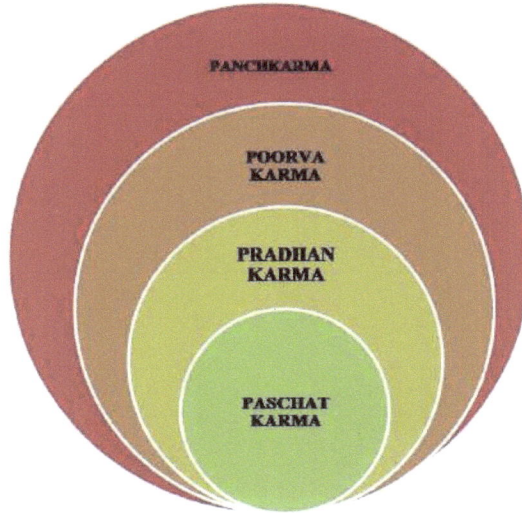

Fig. (15). Three Main procedures of *Panchkarma.*

Poorva Karma (Preparatory Procedure)

It includes carminative (*Deepan*), digestive (*Pachan*), oleation (*Snehan*), and medicated sudation (*Swedan)*. These are beneficial for lubricating; liquefying toxic waste products/metabolites accumulated in various channels of the body, and also helps for easy elimination from the body through the nearest route.

Pradhan Karma (Main Operative Procedure)

After *Poorva Karma* (preparatory procedure), as per requirement the *Pradhan Karma* (main operative procedure), *i.e.*, therapeutic emesis (*Vamana Karma*) and therapeutic Purgation (*Virechan Karma*) to be done, then one should follow medicated Enema (*Vasti Karma*) and medicated nasal drops (*Nasya Karma*).

Paschat Karma (Post-operative Regimen)

After every process of *Pradhan Karma* (main operative procedure), one should follow a special dietary regimen called *Samsarjan Karma* along with *Rasayana* drugs and lifestyle regimen. It is essential to restore the normalcy of body tissue and system as well as helps to rejuvenate the person.

Studies Conducted in Different Panchkarma Procedures

However, in the enlisted *Panchkarma* procedures mentioned above, (Table **8**), maximum research works were conducted on the two procedures which have a significant effect on the management of NDDs and their clinical findings.

Table 8. Common *Panchkarma* procedures used in the treatment of NDDs and their associated symptoms.

S.No.	Name of Panchkarma Procedures	Description	Therapeutic Uses
1.	*MurdhaTaila*	Medicated oil put on the scalp	Insomnia, stress, cerebral atrophy, cerebral ataxia, parkinsonism
2.	*Picchu Dharan*	Cotton socked with medicated oil and put on the scalp for 2-3 hours or more time	Cranial neuropathy
3.	*Siro-Dhara*	Medicated oil, butter, or Ghee poured over forehead in a continuous stream at a specific distance	Insomnia, stress, anxiety neurosis, cranial neuropathy, Parkinson, Alzheimer
4.	*Siro-Vasti*	A unique cap covered the whole head and filled with medicated oil for a specific time duration	Cerebral palsy, cerebral ataxia, Parkinsonism
5.	*Udavartana*	Medicated powder drugs rubbed on specific parts of the body	Hemiplegia, Paraplegia
6.	*Swedana*	Medicated sudation therapy	For associated symptoms of NDDs like stiffness, contractures, myelopathy, and neuropathy, *etc.*
7.	*Vasti*	Medicated enema	Hemiplegia, Paraplegia, Parkinsonism

Shiro-Dhara

Shirodhara is a significant beneficial measure in the Ayurveda system of medicine, which has got worldwide attractiveness because of its simple administration and effectiveness in several neurological, NDDs, and lifestyle diseases. It is a purifying and rejuvenating therapy that eliminates toxins and mental exhaustion as well as relieves stress and any ill effects on the CNS [101].

An open-labeled study finding reveals that *Shirodhara* induced a relaxed state of awareness that resulted in a dynamic psycho-somatic balance in healthy volunteers by monitoring the rating of mood and levels of stress, electrocardiogram, electroencephalogram, and selected biochemical markers of stress and was conducted in the human pharmacology laboratory comparing the baseline variables with values after *Shirodhara* which showed significant improvement in mood scores and the level of stress and significant decrease in the rate of breathing and reduction in diastolic blood pressure along with a reduction in heart rate [102].

Vasti

The method of treatment by which medicated oil or other liquids are administered to the patient through anorectal, urethral, or vaginal route [103].

Vasti is a chief procedure for the management of PD, a degenerative neurological disorder of CNS, mainly affecting the motor system. It is the most common extrapyramidal crippling disease with a prevalence of 1% of the total population. Based on the sign and symptoms like *Kampa* (Tremor), *Stambha* (Rigidity), *Chestasanga* (Bradykinesia, and Akinesia), *Vakvikriti* (disturbance in speech), *etc.*, it can be correlated with *Kampavata* described in Ayurvedic Classics. As in Ayurvedic classics, syndromes are described rather than disease and therefore following the pattern *Kampavata* is not described as a separate entity and described as a type of *VataVyadi* that denotes that this symptom appears when there is a defect in the nervous system and may be accompanied with many other symptoms according to the involvement of different component of the nervous system [104]. Study findings suggested *Tritiya Baladi YapanaVasti* has a significant effect of pacifying the clinical symptoms of PD. This *Vasti* contains *Mucuna pruriens (Kapikachhu),* which is a natural source of Dopamine. Due to decreased dopamine levels in the brain, tremors are mainly found in Parkinson's disease. Thus this *Vasti*may provides relief in the symptoms of tremor [105]. Another study reveals that a 63-year-old female patient presented with complaints of resting tremors in the upper limb (pin–roll type) and head tremors, slow, limited movement, difficulty with walking and balance, insomnia, depression, and face appearing without expression and impairment in activities of daily life, *etc.*, had given *Panchkarma* therapies like *Abhyanga (Dashmool Tail), Svedana (DashmoolaKwatha), Shirobasti (KsheerbalaTaila), Nasya (KsheerbalaTaila), Shiropicchu (KsheerbalaTaila), MustadiYapanaVasti*. Significant improvement was found with consequent treatment for four months along with oral medicines. The assessment was done based on signs and symptoms, bradykinesia, and functional activities [106].

Yoga & Asanas (Yogic Bodily Postures)

Yoga is a unique concept in Ayurveda, which is very useful in NDDs along with *Asanas,* which is helpful in restoring the reminiscence intellect and strength of the patient [107]. Yoga is very aged long-established science that include Yogistic bodily postures (*Asanas*), Yogistic respiratory practices (*Pranayama*), and meditation & recreation techniques (*Dhyana*), exceedingly acknowledged as knowledge rather than philosophy [108]. The center of consideration in Yoga therapy is to bringing balance by all levels of subsistence through a variety of techniques. These *Asanas* (Physical level stimulation), *Pranayama* (Vital energy level stimulation), and meditation (Psychological and intellectual level stimulation) are combinedly known as Yoga [109]. These postures in Yoga helps in blood circulation to vital organs of the body, including the brain, heart, and liver, and check the entry of various types of illness in the human body [110]. Various types of *Asanas* are mentioned in the Yogic test book of *Acharyas Patanjali* which help in the proper functioning of the brain and regenerate the neurons and give strength to the body and mind related to NDDs.

Table 9. Different Yogic postures for the management of NDDs.

S. No.	Name of Yogic Postures	English Name	Benefits	Fig. No.	References
1.	*Padmasana*	Lotus pose	Rest of the mind and calms the brain Stimulates the chakras in the body and boosts consciousness	16	[111]
2.	*Sirsasana*	Headstand pose	Immediately calms the body & stimulates the pituitary gland It stimulates excess blood flow to the brain, which reduces the stress and recovers neuronal activity	17	[111]
3.	*ArdhaMatsyendrasana*	Half spinal twist pose	Relieves stiffness in the back and improves digestion This pose increases the supply of oxygen to the lungs and detoxifies the internal organs	18	[111]
4.	*Shashankasana*	Hare pose	Excellent yoga asana for brain& produces calm, relaxed of mind and increasing the memory power	19	[111, 112]
5.	*Shavasana*	Corpse pose	Relaxes the whole body, which benefits in the patient suffering from the neurological problem and develops excellent concentration, cures insomnia, calms the mind; improve mental health, and also stimulating blood circulation in the brain	20	[111, 113]
6.	*Uttanasana*	Standing forward bend	Release stress, anxiety, depression, and fatigue and Provides calm to the mind, and soothes the nerves.	21	[111, 114]

(Table 9) cont.....

S. No.	Name of Yogic Postures	English Name	Benefits	Fig. No.	References
7.	*Vajrasana*	Diamond pose	Proper digestion and regular practice eliminate constipation and combats acidity. The pose helps the body to relax and increases blood circulation and also improves the flexibility of the lower body, and tones the muscles	22	[111]
8.	*Halasana*	Plow pose	Releases the strain in the back and enhances the posture to reduce stress and calms the brain	23	[111]
9.	*Paschtimottanasana*	Seated forward bend	Relieves mild depression and stress, gives a good stretch to the shoulders, and activates the neurons. This asana reduces headache, fatigue; cures insomnia, and maintain high blood pressure.	24	[111]
10.	*Mayurasana*	Peacock pose	Improves concentration and coordination between the mind and the body.	25	[111]
11.	*Sarvangasana*	Bridge–bandha pose	Alleviate stress and mild depression; provide calms to the brain and CNS.	26	[111, 115]
12.	*Vrikshasana*	Tree pose	Silence to the mind thus good for those who are facing the problem of depression and anxiety. It increases stamina, concentration, immunity and develops esteem and self-confidence. It calms and relaxes the CNS and stretches the entire body from toes to fingers	27	[111, 116]

A study was done in the USA to assess the effect of yoga to improve motor function in a patient with PD. The PD participants were randomized into a yoga intervention group or a control group of no intervention. Assessment of physical function included motor examination scores from the unified PD rating scale, posture, flexibility and strength, and biomechanical measures of balance and gait that occurred at 3-time points were noted at prior, 6 weeks, and 12 weeks of intervention. An Iyengar *Hatha Yoga* program was tailored for 6o minute session twice a week, and a significant improvement was found in motor UPDRS scores and Berg Balance Scale scores in the yoga group [117]. A case study suggested that Yoga is a promising therapeutic modality for NDDs in which therapeutic yoga protocol for a 61-year-old man suffered from Adrenomy eloneuropathy having symptoms of peripheral neuropathy in his legs and feet, lower back pain, and osteoarthritis. The effect was assessed on a patient's quality of life, agility, balance, and peripheral dexterity and showed significant improvement in all symptoms, including improved flexion of the patient's hips, knees, and ankles, improved propulsion phase of walking, and improvement in the patient's ability to stand and balance without an assistive [118]. Another study was conducted on 20 patients with drug-resistant chronic epilepsy to assess the efficacy of a Yoga meditation protocol for 12 weeks as an adjunctive treatment, and complex partial seizures were assessed at 3, 6, and 12 months of the treatment period. The intervention consisted of a Yogic protocol for 20 minutes twice daily at home and supervised sessions every week for 3 months. The result is significant with 6 patients having ≥ 50% reduction in monthly seizure rate from baseline, and a reduction in seizure frequency was noted in all except 1 patient, which shows that Yoga meditation protocol may become a cost-effective and adverse effect-free adjunctive treatment in patients with drug-resistant epilepsies (Table **9**) [119]. Figs. (**16** - **27**) show different yoga techniques widely utilize to manage NDDs.

  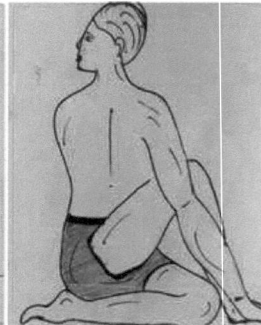

Fig No. 16   Fig No. 17   Fig No. 18

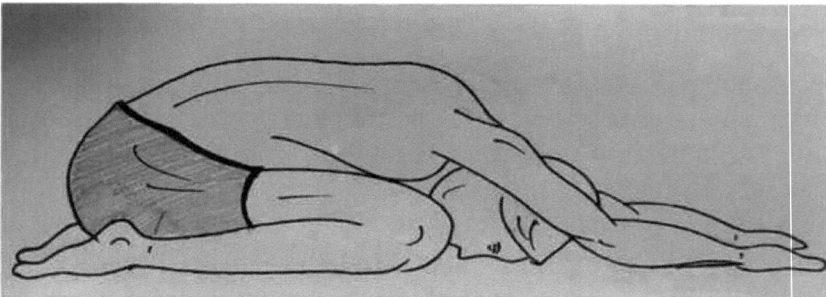

Fig No. 19

(Figs. 16-27) contd.....

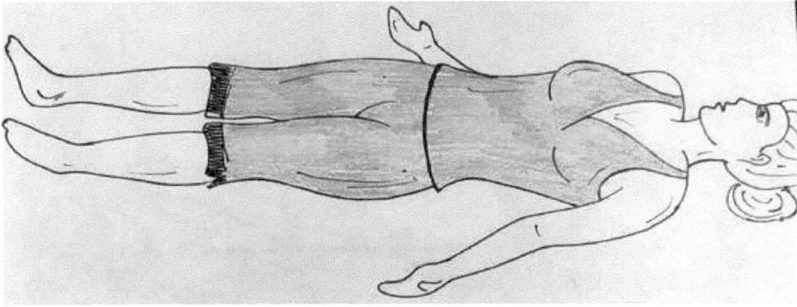

Fig No. 20

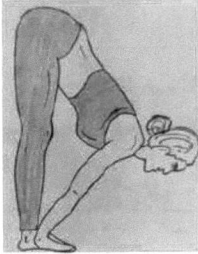  

Fig No. 21　　　　　　Fig No. 22　　　　　　Fig No. 23

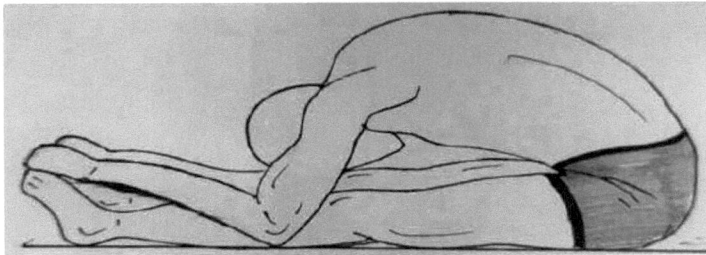

Fig No. 24

Fig No. 25

(Figs. 16-27) contd.....

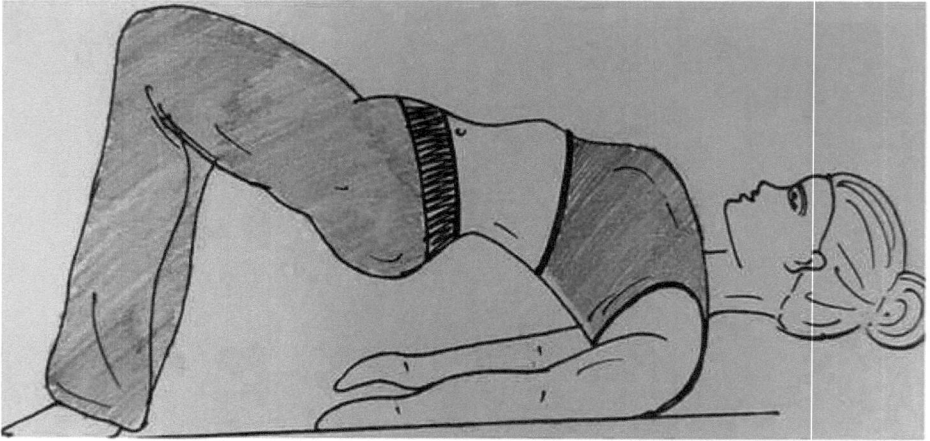

Fig No. 26

Fig No. 27

Figs. (16-27). Different yoga techniques to manage NDDs

CONCLUDING REMARKS

When the necessitate for original and valuable treatments increases, researchers have turned towards ancient knowledge and traditional practices. By using facts based comprehension to investigate the compounds used in the practice of Ayurvedic medicines, various scientists discovered viable alternate treatments that show promising learning ability, anti-inflammatory, reminiscence improving, and neuroprotective effects. Well accepted Ayurvedic medicinal plants (*Ashwagandha, Turmeric, Brahmi, Shankhpushpi*) not only reduce brain aging and persuade antistress and recall enhancing effects which help in regeneration of neural tissues, but also induce antioxidant, anti-inflammatory,anti-amyloidogenic, nutritional, and immune-supportive effects in the body [120]. Apart from studying the pharmacological action of the dynamic components present in the herbs, the rebuilding of Ayurvedic *Panchkarma*therapy in the present era should transcend from the existing methodology of disease detection, prognosis, management, and follow-up to a state of exploring Ayurvedic lifestyles, regimens, and Yogic practices as a part of the treatment regimen. Scientific justification and the certification of Ayurvedic medicines are essential for their dominance evaluation and global acknowledgment. After millennia of safe and efficacious traditional use in brain and nerve disorders, with support of more than a decades-old *in-vitro* and animal studies reports confirming efficacy in several indications and considerable but rigorous test-worthy safety profiles, it is quite apparent that Ayurveda based plant products must pass through the gateways of stringent multicentric unbiased long term human trials in prior being used beyond doubt in the field of neuro medicine and other fields of healthcare [121].

CONSENT FOR PUBLICATION

Not applicable.

CONFLICT OF INTEREST

The authors declare no conflict of interest, financial or otherwise.

ACKNOWLEDGEMENTS

I am thankful to Dr. Rahul Katkar, Junior Resident, Dept of Kaumarbhritya, Institute of Medical Sciences, Banaras Hindu University for providing the original pictures of medicinal plants from the herbal garden, Faculty of Ayurveda, BHU Varanasi.

REFERENCES

[1]   JPND Research. What is Neurodegenerative disease? EU Joint Programme; https://www.neuro degenerationresearch.eu/what/.

[2]   Golde TE, Miller VM. Proteinopathy-induced neuronal senescence: a hypothesis for brain failure in Alzheimer's and other neurodegenerative diseases. Alzheimers Res Ther 2009; 1(2): 5.
[http://dx.doi.org/10.1186/alzrt5] [PMID: 19822029]

[3]   Herczenik E, Gebbink MF. Molecular and cellular aspects of protein misfolding and disease. FASEB J 2008; 22(7): 2115-33.
[http://dx.doi.org/10.1096/fj.07-099671] [PMID: 18303094]

[4]   Jenner P, Olanow CW. Understanding cell death in Parkinson's disease. Ann Neurol 1998; 44(3) (Suppl. 1): S72-84.
[http://dx.doi.org/10.1002/ana.410440712] [PMID: 9749577]

[5]   Adler CH, Connor DJ, Hentz JG, *et al.* Incidental Lewy body disease: clinical comparison to a control cohort. Mov Disord 2010; 25(5): 642-6.
[http://dx.doi.org/10.1002/mds.22971] [PMID: 20175211]

[6]   Gibb WR, Lees AJ. The relevance of the Lewy body to the pathogenesis of idiopathic Parkinson's disease. J Neurol Neurosurg Psychiatry 1988; 51(6): 745-52.
[http://dx.doi.org/10.1136/jnnp.51.6.745] [PMID: 2841426]

[7]   Armstrong RA, Lantos PL, Cairns NJ. Overlap between neurodegenerative disorders. Neuropathology 2005; 25(2): 111-24.
[http://dx.doi.org/10.1111/j.1440-1789.2005.00605.x] [PMID: 15875904]

[8]   Neurodegenerative disease, programme description, National Institute of Environmental Health Sciences 10 Sep 2019.

[9]   Bang J, Spina S, Miller BL. Frontotemporal dementia. Lancet 2015; 386(10004): 1672-82.
[http://dx.doi.org/10.1016/S0140-6736(15)00461-4] [PMID: 26595641]

[10]   Beach TG, Monsell SE, Phillips LE, Kukull W. Accuracy of the clinical diagnosis of Alzheimer disease at national institute on aging Alzheimer disease centers, 2005–2010. J Neuropathol Exp Neurol 2012; 71(4): 266-73.
[http://dx.doi.org/10.1097/NEN.0b013e31824b211b] [PMID: 22437338]

[11]   Ferri CP, Prince M, Brayne C, *et al.* Alzheimer's Disease International. Global prevalence of dementia: a Delphi consensus study. Lancet 2005; 366(9503): 2112-7.
[http://dx.doi.org/10.1016/S0140-6736(05)67889-0] [PMID: 16360788]

[12]   Cornutiu G. G. The epidemiological scale of Alzheimer's disease. J Clin Med Res 2015; 7(9): 657-66.
[http://dx.doi.org/10.14740/jocmr2106w] [PMID: 26251678]

[13]   Blennow K, de Leon MJ, Zetterberg H. Alzheimer's disease. Lancet 2006; 368(9533): 387-403.
[http://dx.doi.org/10.1016/S0140-6736(06)69113-7] [PMID: 16876668]

[14]   Samii A, Nutt JG, Ransom BR. Parkinson's disease. Lancet 2004; 363(9423): 1783-93.
[http://dx.doi.org/10.1016/S0140-6736(04)16305-8] [PMID: 15172778]

[15]   Harper PS. The epidemiology of Huntington's disease. Hum Genet 1992; 89(4): 365-76.
[http://dx.doi.org/10.1007/BF00194305] [PMID: 1535611]

[16]   Dugger BN, Dickson DW. Dickson pathology of neurodegenerative diseases. Cold Spring Harb Perspect Biol 2017; 9(7): a028035.
[http://dx.doi.org/10.1101/cshperspect.a028035] [PMID: 28062563]

[17]   Amador-Ortiz C, Lin WL, Ahmed Z, *et al.* TDP-43 immunoreactivity in hippocampal sclerosis and Alzheimer's disease. Ann Neurol 2007; 61(5): 435-45.
[http://dx.doi.org/10.1002/ana.21154] [PMID: 17469117]

[18]   Hardy J, Selkoe DJ. The amyloid hypothesis of Alzheimer's disease: progress and problems on the road to therapeutics. Science 2002; 297(5580): 353-6.
[http://dx.doi.org/10.1126/science.1072994] [PMID: 12130773]

[19]   Dawson TM, Dawson VL. Molecular pathways of neurodegeneration in Parkinson's disease. Science 2003; 302(5646): 819-22.
[http://dx.doi.org/10.1126/science.1087753] [PMID: 14593166]

[20]   Watt D. Huntington's disease: A disorder of families. Psychol Med 1990; 20(3): 728-31.
[http://dx.doi.org/10.1017/S0033291700017281]

[21]   Rosen DR, Siddique T, Patterson D, *et al.* Mutations in Cu/Zn superoxide dismutase gene are associated with familial amyotrophic lateral sclerosis. Nature 1993; 362(6415): 59-62.
[http://dx.doi.org/10.1038/362059a0] [PMID: 8446170]

[22]   Hsiao KK, Cass C, Schellenberg GD, Bird T. Devine-GageE, WisniewskiH, PrusinerSB. A prion protein variant in a family with the telencephalic form of GerstmannStrausslerScheinkersyndrome. Neurology 1991; 41: 681-4.
[http://dx.doi.org/10.1212/WNL.41.5.681] [PMID: 1674116]

[23]   Paulson HL, Shakkottai VG, Clark HB, Orr HT. Polyglutamine spinocerebellar ataxias - from genes to potential treatments. Nat Rev Neurosci 2017; 18(10): 613-26.
[http://dx.doi.org/10.1038/nrn.2017.92] [PMID: 28855740]

[24]   Hoffmann J. On chronic spinal muscular atrophy in childhood, with a familial basis. Dtsch Z Nervenheilkd 1893; 3: 427-70.
[http://dx.doi.org/10.1007/BF01668496]

[25]   Matute-Blanch C, Montalban X, Comabella M. Multiple sclerosis, and other demyelinating and autoimmune inflammatory diseases of the central nervous system. Handb Clin Neurol 2017; 146: 67-84.
[http://dx.doi.org/10.1016/B978-0-12-804279-3.00005-8] [PMID: 29110780]

[26]   Alafuzoff I, Ince PG, Arzberger T, *et al.* Staging/typing of Lewy body related synuclein pathology. Acta Neuropathol 2009; 117: 635-52.
[http://dx.doi.org/10.1007/s00401-009-0523-2] [PMID: 19330340]

[27]   Alafuzoff I, Pikkarainen M, Neumann M, *et al.* Neuropathological assessments of the pathology in frontotemporal lobar degeneration with TDP43-positive inclusions: an inter-laboratory study by the BrainNet Europe consortium. J Neural Transm (Vienna) 2015; 122(7): 957-72.
[http://dx.doi.org/10.1007/s00702-014-1304-1] [PMID: 25239189]

[28]   Ramanan VK, Saykin AJ. Pathways to neurodegeneration: mechanistic insights from GWAS in Alzheimer's disease, Parkinson's disease, and related disorders. Am J Neurodegener Dis 2013; 2(3): 145-75.
[PMID: 24093081]

[29]   Khoo TK, Yarnall AJ, Duncan GW, *et al.* The spectrum of nonmotor symptoms in early Parkinson disease. Neurology 2013; 80(3): 276-81.
[http://dx.doi.org/10.1212/WNL.0b013e31827deb74] [PMID: 23319473]

[30]   Noyce AJ, Bestwick JP, Silveira-Moriyama L, *et al.* Meta-analysis of early nonmotor features and risk factors for Parkinson disease. Ann Neurol 2012; 72(6): 893-901.
[http://dx.doi.org/10.1002/ana.23687] [PMID: 23071076]

[31]   Bokde ALW, Meaney JFM, Sheehy NP, Reilly RB, Abrahams S, Doherty CP. Advances in diagnostics for neurodegenerative disorders 2011; 3: 17-42.
[http://dx.doi.org/10.1007/978-1-84996-011-3_2]

[32]   Chandra V, Pandav R, Dodge HH, *et al.* Incidence of Alzheimer's disease in a rural community in India: the Indo-US study. Neurology 2001; 57(6): 985-9.
[http://dx.doi.org/10.1212/WNL.57.6.985] [PMID: 11571321]

[33]   Bowirrat A, Friedland RP, Farrer L, Baldwin C, Korczyn A. Genetic and environmental risk factors for Alzheimer's disease in Israeli Arabs. J Mol Neurosci 2002; 19(1-2): 239-45.
[http://dx.doi.org/10.1007/s12031-002-0040-4] [PMID: 12212789]

[34]   Chen JF, Xu K, Petzer JP, *et al.* Neuroprotection by caffeine and A(2A) adenosine receptor inactivation in a model of Parkinson's disease. J Neurosci 2001; 21(10): RC143.
[http://dx.doi.org/10.1523/JNEUROSCI.21-10-j0001.2001] [PMID: 11319241]

[35]   Pogačić Kramp V, Matthisson M, Herrling P. List of drugs in development for neurodegenerative diseases: update October 2011. Neurodegener Dis 2012; 9(4): 210-83.
[http://dx.doi.org/10.1159/000335520] [PMID: 22222285]

[36]   Savitz SI, Fisher M. Future of neuroprotection for acute stroke: in the aftermath of the SAINT trials. Ann Neurol 2007; 61(5): 396-402.
[http://dx.doi.org/10.1002/ana.21127] [PMID: 17420989]

[37]   Cho GW, Noh MY, Kim HY, Koh SH, Kim KS, Kim SH. Bone marrow-derived stromal cells from amyotrophic lateral sclerosis patients have diminished stem cell capacity. Stem Cells Dev 2010; 19(7): 1035-42.
[http://dx.doi.org/10.1089/scd.2009.0453] [PMID: 20030561]

[38]   Orhan IE. *Centella asiatica* (L.) urban: From traditional medicine to modern medicine with neuroprotective Potential. Evid Based Complement Alternat Med 2012; 2012: 946259.
[http://dx.doi.org/10.1155/2012/946259] [PMID: 22666298]

[39]   Shrivastava SR, Shrivastava PS, Ramasamy J. Mainstreaming of Ayurveda, Yoga, Naturopathy, Unani, Siddha, and Homeopathy with the health care delivery system in India. J Tradit Complement Med 2015; 5(2): 116-8.
[http://dx.doi.org/10.1016/j.jtcme.2014.11.002] [PMID: 26151021]

[40]   Behl C. Alzheimer's disease and oxidative stress: implications for novel therapeutic approaches. Prog Neurobiol 1999; 57(3): 301-23.
[http://dx.doi.org/10.1016/S0301-0082(98)00055-0] [PMID: 10096843]

[41]   Aruoma OI, Bahorun T, Jen LS. Neuroprotection by bioactive components in medicinal and food plant extracts. Mutation Research 2003; 544(2): 203-15.
[http://dx.doi.org/10.1016/j.mrrev.2003.06.017]

[42]   Sastri K, Chaturvedi G. Charaka Samhita VidyotiniHindivyakhya Part-1 and Part-2. Chaukhambha Bharati Academy1999 14th Ed.

[43]   Perry EK, Pickering AT, Wang WW, Houghton P, Perry NS. Medicinal plants and Alzheimer's disease: Integrating ethnobotanical and contemporary scientific evidence. J Altern Complement Med 1998; 4(4): 419-28.
[http://dx.doi.org/10.1089/acm.1998.4.419] [PMID: 9884179]

[44]   Jeyam M, Karthika R, Poornima V, Sharanya M. Molecular understanding and *in silico* validation of traditional medicines for Parkinson's disease. Asian Jour of Pharma and Clinical Res 2012; 5(4): 125-8.

[45]   Pandey SK, Jangra MK, Yadav AK. Herbal and synthetic approaches for the treatment of epilepsy. Intern J NutrPharmacol Neurol Dis 2014; 4: 43-52.
[http://dx.doi.org/10.4103/2231-0738.124613]

[46]   Rasool M, Malik A, Qureshi MS, Manan A, Pushparaj PN. Muhammad Asif. Recent updates in the treatment of neurodegenerative disorders using natural compounds. Evid Based Complement Alternat Med 2014; 979730: 1-7.

[47]   Rao RV, Descamps O, John V, Bredesen DE. Ayurvedic medicinal plants for Alzheimer's disease: a review. Alzheimers Res Ther 2012; 4(3): 22.
[http://dx.doi.org/10.1186/alzrt125] [PMID: 22747839]

[48]   Hussain I, Khan N, Khanetal H. Screening of anti-oxidant activities of selected medicinal plants. World Appl Sci J 2010; 11(3): 338-40.

[49]   Bagchi P, Kar A, Vinobha SC. Establishing an in-silico Ayurvedic medication towards treatment of Schizophrenia. Intern Journ of Syst Biol 2009; 1(2): 46-50.

[50]   Bhattacharya SK, Muruganandam AV. Adaptogenic activity of Withania somnifera: an experimental study using a rat model of chronic stress. Pharmacol Biochem Behav 2003; 75(3): 547-55.

[http://dx.doi.org/10.1016/S0091-3057(03)00110-2] [PMID: 12895672]

[51]   G V, K SP, v L, Rajendra W. The antiepileptic effect of *Centella asiatica* on the activities of Na/K, Mg and Ca-ATPases in rat brain during pentylenetetrazol-induced epilepsy. Indian J Pharmacol 2010; 42(2): 82-6.
[PMID: 20711371]

[52]   Bairy KL, Rao Y, Kumar KB. Efficacy of *Tinospora cordifolia* on learning and memory in healthy volunteers: A double-blind, randomized, placebo controlled study. Iran Journ of Pharm &Therap 2004; 3: 57-60.

[53]   Agarwa P, Sharma B, Fatima A, Jain SK. An update on Ayurvedic herb Convolvulus pluricaulis Choisy. Asian Pac J Trop Biomed 2014; 4(3): 245-52.
[http://dx.doi.org/10.1016/S2221-1691(14)60240-9] [PMID: 25182446]

[54]   Hassanien MF, Kinni SG, Moersel JT. Bioactive lipids, fatty acids and radical scavenging activity of Indian Celastruspaniculatus oil. J Appl Bot Food Qual 2010; 83(2): 157-62.

[55]   Nagashayana N, Sankarankutty P, Nampoothiri MR, Mohan PK, Mohanakumar KP. Association of L-DOPA with recovery following Ayurveda medication in Parkinson's disease. J Neurol Sci 2000; 176(2): 124-7.
[http://dx.doi.org/10.1016/S0022-510X(00)00329-4] [PMID: 10930594]

[56]   Nam SM, Choi JH, Yoo DY, *et al.* Effects of curcumin (Curcuma longa) on learning and spatial memory as well as cell proliferation and neuroblast differentiation in adult and aged mice by upregulating brain-derived neurotrophic factor and CREB signaling. J Med Food 2014; 17(6): 641-9.
[http://dx.doi.org/10.1089/jmf.2013.2965] [PMID: 24712702]

[57]   Das A, Shanker G, Nath C, Pal R, Singh S, Singh H. A comparative study in rodents of standardized extracts of *Bacopa monniera* and Ginkgo biloba: anticholinesterase and cognitive enhancing activities. Pharmacol Biochem Behav 2002; 73(4): 893-900.
[http://dx.doi.org/10.1016/S0091-3057(02)00940-1] [PMID: 12213536]

[58]   Ahmed F, Chandra NS, Urooj A, Rangappa KS. *In vitro* antioxidant and anti-cholinesterase activity of Acorus calamus and Nardostachysjatamansi rhizomes. J Pharm Res 2009; 2(5): 830-3.

[59]   Joy PP, Thomas J, Mathew S, Skaria BP. (2001) Medicinal Plants. In: Bose, T.K., Kabir, J., Das, P. and Joy, P.P., Eds., Tropical Horticulture, Naya Prokash, Calcutta, 449-632.

[60]   Auddy B, Hazra J, Mitra A. A standardized Withania somnifera extract significantly reduces stress related parameters in chronically stressed humans: A double–blind randomized placebo controlled study. JANA 2008; 11(1): 50-6.

[61]   Jayaprakasam B, Padmanabhan K, Nair MG. Withanamides in Withania somnifera fruit protect PC-12 cells from beta-amyloid responsible for Alzheimer's disease. Phytother Res 2010; 24(6): 859-63.
[http://dx.doi.org/10.1002/ptr.3033] [PMID: 19957250]

[62]   Birla H, Keswani C, Rai SN, *et al.* Neuroprotective effects of Withania somnifera in BPA induced-cognitive dysfunction and oxidative stress in mice. Behav Brain Funct 2019; 15(1): 9.
[http://dx.doi.org/10.1186/s12993-019-0160-4] [PMID: 31064381]

[63]   Kurapati KR, Atluri VS, Samikkannu T, Nair MP, Madhavan PN. Ashwagandha (Withania somnifera) reverses β-amyloid1-42 induced toxicity in human neuronal cells: implications in HIV-associated neurocognitive disorders (HAND). PLoS One 2013; 8(10): e77624.
[http://dx.doi.org/10.1371/journal.pone.0077624] [PMID: 24147038]

[64]   Chen CL, Tsai WH, Chen CJ, Pan TM. *Centella asiatica* extract protects against amyloid β_{1-40}-induced neurotoxicity in neuronal cells by activating the antioxidative defence system. J Tradit Complement Med 2015; 6(4): 362-9.
[http://dx.doi.org/10.1016/j.jtcme.2015.07.002] [PMID: 27774420]

[65]   Lokanathan Y, Omar N, Ahmad Puzi NN, Saim A, Hj Idrus R. Recent Updates in Neuroprotective and Neuroregenerative Potential of *Centella asiatica*. Malays J Med Sci 2016; 23(1): 4-14.

[PMID: 27540320]

[66]   Siddique YH, Naz F, Jyoti S, *et al*. Effect of *Centella asiatica* leaf extract on the dietary supplementation in transgenic drosophila model of Parkinson's disease. Parkinsons Dis 2014; 2014: 262058.
[http://dx.doi.org/10.1155/2014/262058] [PMID: 25538856]

[67]   Sharma R, Amin H, Prajapati P. Therapeutic vistas of Guduchi (*Tinospora cordifolia*): A medico-historical memoir. J Res Educ Indian Med 2014; 20(2): 113-28.

[68]   Phukan P, Bawari M, Sengupta M. Promising neuroprotective plants from northeast India. Int J Pharm Pharm Sci 2015; 7(3): 28-39.

[69]   Bairy KL, Rao Y, Kumar KB. Efficacy of *Tinospora cordifolia* on learning and memory in healthy volunteers: A double-blind, randomized, placebo controlled study. Iran Journ of Pharma &Therap 2004; 3: 57-60.

[70]   Singh SS, Pandey SC, Srivastava S, Gupta VS, Patro B, Ghosh AC. Chemistry and medicinal properties of *Tinospora cordifolia* (Guduchi). Indian J Pharm 2003; 35(2): 83-91.

[71]   Sharma K, Bhatnagar M, Kulkarni SK. Effect of Convolvulus pluricaulis Choisy and Asparagus racemosus Willd on learning and memory in young and old mice: a comparative evaluation. Indian J Exp Biol 2010; 48(5): 479-85.
[PMID: 20795365]

[72]   Amin H, Sharma R, Vyas H, Vyas M, Prajapati PK, Dwivedi R. Nootropic (medhya) effect of Bhāvita Śaṅkhapuṣpī tablets: A clinical appraisal. Anc Sci Life 2014; 34(2): 109-12.
[http://dx.doi.org/10.4103/0257-7941.153476] [PMID: 25861147]

[73]   Bihaqi SW, Singh AP, Tiwari M. Supplementation of Convolvulus pluricaulis attenuates scopolamine-induced increased tau and amyloid precursor protein (AβPP) expression in rat brain. Indian J Pharmacol 2012; 44(5): 593-8.
[http://dx.doi.org/10.4103/0253-7613.100383] [PMID: 23112420]

[74]   Dubey GP, Pathak SR, Gupta BS. Combined effect of Brahmi (*Bacopa monniera*) and Shankhpushpi (Convolvulus pluricaulis) on cognitive functions. Pharmaco psycho ecol 1994; 7: 249-51.

[75]   Kumar MHV, Gupta YK. Antioxidant property of *Celastrus paniculatus* willd.: a possible mechanism in enhancing cognition. Phytomedicine 2002; 9(4): 302-11.
[http://dx.doi.org/10.1078/0944-7113-00136] [PMID: 12120811]

[76]   Rai SN, Birla H, Zahra W, Singh SS, Singh SP. Immunomodulation of Parkinson's disease using Mucuna pruriens (Mp). J Chem Neuroanat 2017; 85(11): 27-35.
[http://dx.doi.org/10.1016/j.jchemneu.2017.06.005] [PMID: 28642128]

[77]   Rai SN, Birla H, Singh SS, *et al*. *Mucuna pruriens* Protects against MPTP Intoxicated Neuroinflammation in Parkinson's Disease through NF-κB/pAKT Signaling Pathways. Front Aging Neurosci 2017; 9: 421.
[http://dx.doi.org/10.3389/fnagi.2017.00421] [PMID: 29311905]

[78]   Pras N, Woerdenbag HJ, Batterman S, Visser JF, Van Uden W. Mucuna pruriens: improvement of the biotechnological production of the anti-Parkinson drug L-dopa by plant cell selection. Pharm World Sci 1993; 15(6): 263-8.
[http://dx.doi.org/10.1007/BF01871128] [PMID: 8298586]

[79]   Aggarwal BB, Harikumar KB. Potential therapeutic effects of curcumin, the anti-inflammatory agent, against neurodegenerative, cardiovascular, pulmonary, metabolic, autoimmune and neoplastic diseases. Int J Biochem Cell Biol 2009; 41(1): 40-59.
[http://dx.doi.org/10.1016/j.biocel.2008.06.010] [PMID: 18662800]

[80]   Ganguli M, Chandra V, Kamboh MI, *et al*. Apolipoprotein E polymorphism and Alzheimer disease: The IndoUS cross national dementia study. Arch Neurol 2000; 57(6): 824-30.
[http://dx.doi.org/10.1001/archneur.57.6.824] [PMID: 10867779]

[81]   Breitner JC, Welsh KA, Helms MJ, *et al*. Delayed onset of Alzheimer's disease with nonsteroidal anti-inflammatory and histamine H2 blocking drugs. Neurobiol Aging 1995; 16(4): 523-30.
[http://dx.doi.org/10.1016/0197-4580(95)00049-K] [PMID: 8544901]

[82]   Shinomol GK, Muralidhara , Bharath MM. Exploring the role of 'Brahmi' (Bacopa monnieri and *Centella asiatica*) in brain function and therapy. Recent Pat Endocr Metab Immune Drug Discov 2011; 5(1): 33-49.
[http://dx.doi.org/10.2174/187221411794351833] [PMID: 22074576]

[83]   *Bacopa monniera*. Monograph. Altern Med Rev. 2004 Mar;9(1):79-85.
[PMID: 15005647]

[84]   Uabundit N, Wattanathorn J, Mucimapura S, Ingkaninan K. Cognitive enhancement and neuroprotective effects of Bacopa monnieri in Alzheimer's disease model. J Ethnopharmacol 2010; 127(1): 26-31.
[http://dx.doi.org/10.1016/j.jep.2009.09.056] [PMID: 19808086]

[85]   Bhattacharya SK, Bhattacharya A, Kumar A, Ghosal S. Antioxidant activity of *Bacopa monniera* in rat frontal cortex, striatum and hippocampus. Phytother Res 2000; 14(3): 174-9.
[http://dx.doi.org/10.1002/(SICI)1099-1573(200005)14:3<174::AID-PTR624>3.0.CO;2-O]   [PMID: 10815010]

[86]   Akram M, Nawaz A. Effects of medicinal plants on Alzheimer's disease and memory deficits. Neural Regen Res 2017; 12(4): 660-70.
[http://dx.doi.org/10.4103/1673-5374.205108] [PMID: 28553349]

[87]   Gupta A, Singh MP, Sisodia S. A review on herbal Ayurvedic medicinal plants and its association with memory functions. Journal of Phytopharmacology 2018; 7(2): 162-6.

[88]   Rasool M, Malik A, Qureshi MS, *et al*. Recent updates in the treatment of neurodegenerative disorders using natural compounds. Evid Based Complement Alternat Med 2014; 2014: 979730.
[http://dx.doi.org/10.1155/2014/979730] [PMID: 24864161]

[89]   Dwivedi V, Tripathi BK, Mutsuddi M, Lakhotia SC. Ayurvedic AmalakiRasayana and Rasa-Sindoor suppress neurodegeneration in fly models of Huntington's and Alzheimer's diseases. Curr Sci 2013; 105(12): 1711-23.

[90]   Shetty C, Kumary S. Effects of Swarnaprashana on hippocampus of albino rats: an experimental study. Intern Ayur Med Jour 2017; 5(11): 4064-70.

[91]   Yadav KD, Reddy KRC, Kumar V. Beneficial effect of *Brahmi Ghrita* on learning and memory in normal rat. Ayu 2014; 35(3): 325-9.
[http://dx.doi.org/10.4103/0974-8520.153755] [PMID: 26664242]

[92]   Ekka D, Dubey S, Dhruva S. Effect of Rajat Bhasma with Smritisagar Rasa in Parkinson. Journ of Ayur&Integ Med Sci 2017; 2(4): 146-50.
[http://dx.doi.org/10.21760/jaims.v2i4.9341]

[93]   Parekar RR, Jadhav KS, Marathe PA, Rege NN. Effect of *Saraswatarishta* in animal models of behavior despair. J Ayurveda Integr Med 2014; 5(3): 141-7.
[http://dx.doi.org/10.4103/0975-9476.140469] [PMID: 25336844]

[94]   Khory RN, Katraka MN. Materia Medica of India and their therapeutics. Delhi, India: Neeraj Publishing House, 1984; 64.

[95]   Krishnamurthy MS. Aeitology, symptomatology and treatment of vatic disorders. Basavarajeeyam first edition. 2014; 149.

[96]   Hügel HM. Brain food for Alzheimer-free ageing: focus on herbal medicines. Adv Exp Med Biol 2015; 863: 95-116.
[http://dx.doi.org/10.1007/978-3-319-18365-7_5] [PMID: 26092628]

[97]   Shrivastva S. Sharangadhar Smhita, Uttara Khanda, Edition, Chaukhambha Orentalia, Varanasi,

2016; p.450.

[98]   Saxena N. Vangasena Samhita or Cikitsasara Samgraha of Vangasena: v.1 (Sanskrit) 2004; 1: 409.

[99]   Shastri K, Chaturvedi GN. Charaka Samhita: Agnivesh revised by Charaka and Dridhabala, Vidyotini Hindi Commentar. Chaukhambha Sanskrit series 1986. 13th ed.

[100] Shastry RD. Basvarajiyam Basavraj. ,Chaukhambha Sanskrit Series 1987 Chapter 6.

[101] Xu F, Uebaba K, Ogawa H, *et al.* Pharmaco-physio-psychologic effect of Ayurvedic oil-dripping treatment using an essential oil from Lavendula angustifolia. J Altern Complement Med 2008; 14(8): 947-56.
[http://dx.doi.org/10.1089/acm.2008.0240] [PMID: 18990044]

[102] Dhuri KD, Bodhe PV, Vaidya AB. Shirodhara: A psycho-physiological profile in healthy volunteers. J Ayurveda Integr Med 2013; 4(1): 40-4.
[http://dx.doi.org/10.4103/0975-9476.109550] [PMID: 23741161]

[103] Shastry RD. Ashtang Hridayam Sutrasthana 1/25. Janhavi Prakashan Kolhapur 2006, 1st ed.

[104] Sushruta Samhita Chikitsasthan 35/27.Chaukhamba Sanskrit Sansthan, 2018; 192.

[105] Ranajan M, Jaiswal A, Thakar AB. Efficacy of Yapana Basti in the management of Parkinson's disease. Punarnav 2015; 3(2): 1-5.

[106] Verma J, Prakash S. Effect of *Panchkarma* therapy in the management of Kampavata w.s.r to Parkinson's disease - A case study. Int J Adv Res (Indore) 2018; 6(9): 312-8.

[107] Streeter CC, Gerbarg PL, Saper RB, Ciraulo DA, Brown RP. Effects of yoga on the autonomic nervous system, gamma-aminobutyric-acid, and allostasis in epilepsy, depression, and post-traumatic stress disorder. Med Hypotheses 2012; 78(5): 571-9.
[http://dx.doi.org/10.1016/j.mehy.2012.01.021] [PMID: 22365651]

[108] Berger B, Owen D. Yoga and stress reduction and mood enhancement in four exercise modes: swimming, body conditioning, Hatha Yoga and fencing. Res Q Exerc Sport 1988; 59: 148-59.
[http://dx.doi.org/10.1080/02701367.1988.10605493]

[109] Innes KE, Bourguignon C, Taylor AG. Risk indices associated with the insulin resistance syndrome, cardiovascular disease, and possible protection with yoga: a systematic review. J Am Board Fam Pract 2005; 18(6): 491-519.
[http://dx.doi.org/10.3122/jabfm.18.6.491] [PMID: 16322413]

[110] Asha. Yoga Asanas for your. Brain 2014; (July): 23.

[111] Ramya Achanta. Effective Yoga poses to increase your brain power 2018.

[112] Prema Sitham. The procedure and benefits of Shashankasana Yoga Asanas.

[113] Raj Shavasana. Savasana Yoga (Corpse Pose) benefits and steps 2014.

[114] Cnyha. The health benefits of Uttanasana (Standing Forward Bend Pose) 2018.

[115] Namita Nayyar. Counting health benefits of Setu Bandha Sarvangasana 2016.

[116] The art of living. Tree Pose –Vrikshasana 2016.

[117] Yvonne S, Sharma N, Kluding P, *et al.* Effect of Yoga on motor function in people with Parkinson's disease: A randomized, controlled pilot study. Int J Yoga 2015; 8(1): 74-9.
[http://dx.doi.org/10.4103/0973-6131.146070] [PMID: 25558138]

[118] Muhammad CM, Moonaz SH. Yoga as therapy for neurodegenerative disorders: A case report of therapeutic Yoga for Adrenomyeloneuropathy. Integr Med (Encinitas) 2014; 13(3): 33-9.
[PMID: 26770098]

[119] Rajesh B, Jayachandran D, Mohandas G, Radhakrishnan K. A pilot study of a yoga meditation protocol for patients with medically refractory epilepsy. J Altern Complement Med 2006; 12(4): 367-

71.
[http://dx.doi.org/10.1089/acm.2006.12.367] [PMID: 16722786]

[120] Manyam BV. Dementia in Ayurveda. J Altern Complement Med 1999; 5(1): 81-8.
[http://dx.doi.org/10.1089/acm.1999.5.81] [PMID: 10100034]

[121] Tandon M, Prabhakar S, Pandhi P. Pattern of use of complementary/alternative medicine (CAM) in epileptic patients in a tertiary care hospital in India. Pharmacoepidemiol Drug Saf 2002; 11(6): 457-63.
[http://dx.doi.org/10.1002/pds.731] [PMID: 12426930]

CHAPTER 3

Role of Phytochemicals in Neurodegenerative Disorders

Shambhoo Sharan Tripathi[1,*], Raushan Kumar[1], Prabhash Kumar Pandey[1], Abhishek Kumar Singh[2] and Nidhi Gupta[3,4]

[1] *Department of Biochemistry, University of Allahabad, Allahabad-211002, Uttar Pradesh, India*

[2] *Amity Institute of Neuropsychology and Neurosciences, Amity University, Noida-201313, Uttar Pradesh, India*

[3] *Department of Psychology, VBM College (J.P. University, Chapra), Siwan-841226, Bihar, India*

[4] *Department of Psychology, D.D.U. Gorakhpur University, Gorakhpur-273001, Uttar Pradesh, India*

Abstract: Neurodegenerative disorders (NDs) are one of the leading serious problems worldwide, not only for developed countries but also for developing countries. NDs can be described as a progressive loss of neurons of the central nervous system that leads to cognitive impairment in individuals. The generation of excess reactive oxygen species is one of the reasons for the pathogenesis of NDs. From the various study, it has been established that the use of antioxidants may reduce the onset of NDs. The treatment of these diseases is very costly; for example, the cost of AD worldwide is estimated to be~ $800 billion in 2015. Moreover, in 2017 the cost of PD is reported to have been greater than ~$14 billion in the United States. Now, the researchers have focused on the screening of phytochemicals that have a huge antioxidant effect and neuroprotective ability. Phytochemicals are plant-derived biochemical, and they are described to have a protective effect on oxidative stress (OS), inflammation and provide better mental health. In this chapter, we have incorporated some important phytochemicals that have a great capacity to protect our brain cells and slow down or inhibit NDs pathogenesis.

Keywords: Antioxidant, Neurodegenerative disorders, Neuroinflammation, Oxidative stress, Phytochemicals.

INTRODUCTION

Neurodegenerative disorders (NDs) are a vital problem in both developed and developing countries of the world. Neurodegeneration can be characterized by chronic progressive loss of neuronal cells of the central nervous system (CNS)

* **Corresponding author Shambhoo Sharan Tripathi:** Department of Biochemistry, University of Allahabad, Allahabad-211002, India; Tel: +91 8887899227; E-mail: shambhudna@gmail.com

Sachchida Nand Rai (Ed.)

that culminated in functional and mental impairments [1]. Neurodegeneration is a consequence of elevated reactive oxygen/nitrogen species (RO/NS) formation, protein misfolding or aggregation, failure of mitochondrial function, synaptic loss, and apoptosis in neuronal cells [2]. An elevated augmentation of protein aggregation influences the signaling in neuronal cells and is an essential reason for cell death [3]. Some biological agents, like viruses, are also able for neuronal loss and lead to neurodegeneration [4]. Similarly, in multiple sclerosis (MS), the pathological features involve the permeability of the blood-brain barrier (BBB), the destruction of the myelin sheath, damage of the axon, the formation of the glial scar, and the presence of inflammatory cells, mostly lymphocytes infiltrated into the CNS [5, 6].

Aging is also one of the vital causes of neurodegeneration [7]. Most of the aging-associated NDs are distinguished by the diseases-specific misfolded proteins in nerve cells [8, 9]. For example, Alzheimer disease (AD), Parkinson's disease (PD), amyotrophic lateral sclerosis (ALS), and Huntington's disease (HD) are characterized by the aggregation of beta-amyloid and tau/phosphorylated tau, α-synuclein, superoxide dismutase, and mutant huntingtin (Htt) proteins respectively [10].

According to an epidemiologically study, AD constitutes worldwide 60%-80% of all dementia and significant form of memory loss-related disorders, hitting an expected 24 million people globally [11, 12]. PD is on rank second in neurodegenerative disorder, and the prevalence of PD is considered to be 0.3% in the global population, ~1% in aged persons (>60 years), and ~3% in more than 80 years of peoples [13, 14]. The frequency of HD was found to be 4–8 in 100000 people in Europe [15]; there is little information regarding the epidemiology of ALS and MS in the world. There is a higher incidence in men compared to women (1.5:1), with an average age of onset between 58 and 60 years and mean survival of 3 to 4 years after diagnosis [16].

Based on the pathogenesis of NDs, there is a requirement for therapeutic interventions that can protect from the most common hallmark of NDs. Phytochemicals are naturally occurring chemicals that can be the most suitable therapeutic intervention for it. In this book chapter, we tried to cover most of the therapeutic role of phytochemicals in NDs.

PATHOGENESIS OF NEURODEGENERATION

In the pathogenesis of NDs like AD [17], PD [18], HD [19], ALS [20], *etc.*, the involvement of mitochondria is very obvious. The role of mitochondrial dysfunction is consists of respiratory chain malfunction and production of OS reduction in ATP generation, calcium signaling dysregulation, the opening of mitochondrial membrane transition pore, a perturbation in dynamics of mitochondrial, and deregulated mitophagy [21, 22]. Most of the mitochondria's functions are interdependent and can be present together in a different mode in the various disorders [23]. OS is one of the primary causes for NDs and is a state in which the equilibrium between the ROS production and the antioxidant level is significantly disrupted, and as a consequence, cells undergo apoptosis [24, 25]. Excessive ROS production makes several changes in biomolecules function and contributes to the pathogenesis of NDs. ROS alters the biomolecules, *i.e.*, protein, lipid, DNA, and RNA of the cells [26]. These changes are very useful biomarkers for the detection of NDs. The polyunsaturated fatty acids (*e.g.*, arachidonic acids and docosahexaenoic acids) are present in a sufficient amount in the brain, and due to ROS attack, they change into oxidized form (malondialdehyde and 4-hydroxynonenal). ROS converts protein through oxidizing both the backbone and the side chain into carbonyl form. Uncontrolled oxidation of lipid in the presence of excess OS promotes protein aggregation in the pathogenesis of NDs [27, 28]. During the pathogenesis of NDs, changes in protein due to OS are protein carbonylation, nitration, and chlorination that leads to change in protein structure and functions [29]. ROS also engages in the conversion of nucleic acids in several ways, causing DNA-protein crosslinks, breaks in the strand, and modifies purine and pyridine bases resulting in DNA mutations [30, 31].

AD is the most prevalent neurodegenerative disease, distinguished by gradual neuronal degeneration linked with the aggregation of extracellular amyloid (Aβ) protein, and intracellular tau tangles. AD brains are associated with ROS mediated-damage; there is an increase in levels of lipid peroxidation in the brain and cerebrospinal fluid of AD patients compared to healthy controls [32]. Moreover, the concentration of protein carbonyl is increased in the AD brain [29, 33]. There is also an increment in ROS-induced hydroxylated guanine in AD samples compared to controls [34].

After AD, PD is the second most prevalent neurodegenerative disease. It is characterized by a gradual decline of dopaminergic neurons in the substantia nigra (SN) and aggregation of the protein α-synuclein. In the PD brain, the concentration of the markers of lipid peroxidation increases significantly in the SN region of the brain [35, 36]. The presence of excessive protein carbonyl in the PD brain also signifies the ROS-induced injury [29, 37, 38], and there is some proof to recommend a function for nitration and nitrosylation of specific proteins due to RNS in the PD brain [39, 40].

ROLE OF PHYTOCHEMICALS IN NEURODEGENERATION

In the present scenario, where the incidence of NDs is increasing day by day to which the OS contributes significantly. The researchers have focused on the screening of phytochemicals that would have a huge antioxidant effect and neuroprotective ability. Phytochemicals are plant-derived biochemicals, and they are explained to have a protective effect on OS and neuroinflammation, which are the important hallmarks of NDs [41, 42]. Curcumin, quercetin, diallyltrisulfide, flavonoids, and epigallocatechin-3-gallate (EGCG) are traditional phytochemicals and the part of our daily lives. Phytochemicals are an excellent immune system booster, attenuate platelet aggregation, and improve hormone metabolism [43, 44]. The majority of phytochemicals have an antioxidant role and target the stress response pathways which the cells use as protection[44]. Furthermore, the phytochemicals might also act as a ligand for specific receptors present on the cells or nuclei, thereby playing a vital role in a downstream signaling pathway and regulating the gene expression of antioxidant proteins [45, 46].

NERUROPROTECTING PHYTOCHEMICALS

It has been observed that phytochemicals are in use for protection from NDs in therapeutic interest. Several phytochemicals are very useful in animal and cell-based models of neurological disorders. In which, chalcone is an example for neuroprotection toward ischemic brain injury, and another phytochemical called piceatannol gives protection to neuronal cell line from Aβ- induced apoptosis [47]. A large human study related to NDs, have furnished evidence that phytochemicals derived from fruits and vegetables have neuro protecting activity [48]. From the data of different research groups, it is clear that most of the phytochemicals have active chemical ingredients that possess antioxidant activity [43, 44]. These active ingredients target OS and exhibit neuroprotective effects, *e.g.*, green tea, resveratrol, quercetin, and catechins diminished OS and protect hippocampal neurons in-*vitro* against nitric oxide-induced cell death [48, 49]. Several articles have been published reporting neuroprotective effects of compounds in natural products, including a-tocopherol, lycopene, resveratrol, and ginsenosides. Some of the important phytochemicals in neuroprotection have given below (Table **1**).

Table 1. List of neuroprotective phytochemicals.

S.no	Phytochemicals	Source	Mode of Action	References
1.	**Epigallocatechin-3-Galate-**	*Camellia sinensis*	Inhibiting ERK and NF-kB α-secretase, decrease ROS level.	[58]

(Table 1) cont.....

2.	**Berberine**	*Coptischinensis berberine*	Activates PI3K/Akt/Nrf2 pathways,decreased ROS level, inhibits Cox2,TNF-α,&NF-κB.	[71]
3.	**Curcumin-**	*Curcuma longa*	Inhibits Cas3, TNF-α, andNF-κB levels, reduced ROS level.	[143]
4.	**Resveratrol**	Found in grapes, peanuts, wine, and tea	Inhibiting TNF-α and IL-1beta levels, activates ERK1-2/CREB signaling pathways, activates, AMPK-SIRT--autophagy pathway.	[144]
5.	**Quercetin**	Apples, Berries, Ginkgo biloba	Quercetin inhibits inflammatory enzymes cyclooxygenase (COX) and lipooxygenase, reduced ROS level,	[145]
6.	**Fisetin**	Apples, Strawberries	Suppressed the ASK-1, p-JNK, and p53 expression proteins. Inhibits TNF-α, and NF-κB levels reduced ROS level.	[97]
7.	**Ginsenosides**	Components of ginseng	Hydroxyl radical removing activity activates Nrf2 pathways, contains SOD-like activity.	[146]
8.	**Sulforaphane (SFN)**	Derived from cruciferous vegetables	Reduced levels of IFN-γ, MCP-1, and TNF-α, and increased IL-10 levels	[147]
9.	**Genistein**	Phytoestrogen from soy isoflavonoid	Angiogenesis inhibitors, significantly decreased reactive oxygen species levels and induced the expression of the antioxidant enzymes manganese (Mn) superoxide dismutase (SOD) and catalase, which were associated with AMP-activated protein kinase (AMPK)	[148]
10	**Allium and allicin**	Found in garlic and onions	Scavenge hydroxyl radicals, prevented the lipid peroxidation, scavenge Oxygen free radicals,activates transient receptor potential (TRP) ion channels in the plasma membrane of neurons	[149]
11.	**Baicalein**	*Scutellaria baicalensis* Georgi	Activates Nrf2/HO-1 signaling pathway, prevented the lipid peroxidation, scavenge Oxygen free radicals.	[150]
12	**Icariin**	*Epimedium brevicornum* Maxim	Activation of the PI3K/Akt signaling pathways that induces the inhibition of GSK-3β and, consequently, reduces tau protein hyperphosphorylation	[134]
13	**Morin**	Moraceae family	Antioxidant, anti-inflammatory, and antiproliferative effects.	[138]

Epigallocatechin-3-Gallate

Epigallocatechin-3-gallate (EGCG) is a phytochemical and active ingredient of green tea leaves [50]. In the last few decades, it is in use as a potential dietary component for delaying neurodegeneration [51, 52]. There are lots of studies that have been done to find out the effect of EGCG on AD animal models. In one study, it has been seen that EGCG treatment in the AD mouse model drastically diminished amyloid plaques. The underlying mechanism of the removal of plaque is the activation of the nonamyloidogenic α-secretase proteolytic pathway [50, 53]. Moreover, EGCG overcomes Aβ-induced cellular toxicity by diminishing ROS-induced NF-κB signaling and mitogen-activated protein kinase (MAPK) activation as well as c-Jun N-terminal kinase (JNK) and p38 signalling [50]. In another study, it has been proposed that adults who drink three or more cups of green tea daily have lowered the risk of developing PD [54]. It also reported that the free radical scavenging mechanism of EGCG is due to increasing cellular glutathione, which further initiates CREB (cAMP response element-binding protein) and Bcl-2 (B-cell lymphoma 2), which leads to positive manifestations [55]. PD is characterized by the progressive loss of tyrosine hydroxylase (TH) positive cells. Therefore, one study has advised that the concomitant consumption of EGCG inhibits the death of TH cells in the SN region of the brain [56]. From these reports, it is clear that EGCG can be practiced as a remedial for NDs. EGCG exerts neuroprotection by inhibiting dopamine (DA) uptake, which in turn stops neuronal cell death [57]. The underlying mechanism of this process is that it acts on the enzyme catechol-O-methyltransferase (COMT), which prevents the metabolism of DA. Furthermore, EGCG is considered to improve the proteolytic breakdown of α-synuclein, which is get aggregated in PD, transforming it into a less toxic form [57, 58]. On the other side, EGCG exerts its effect in a concentration-dependent manner. For neuroprotection, low concentrations of EGCG required while high concentrations have antiproliferative effects and stop angiogenesis [57]. Due to the presence of 3, 4, 5trihydroxy B ring and belongs to the catechol family, EGCG displays antioxidant effects and alleviates the free radicals [59]. Nrf2/ARE signaling pathway involves the EGCG mediated antioxidant effect [60]. The active phytochemical of green tea (*i.e.,* EGCG) also very effective in ameliorating Htt-induced toxicity in the HD model [61, 62].

Berberine

From different studies, this has been established that Berberine provides neuroprotection by monitoring neurotrophin levels in various neurodegenerative models [63, 64]. In the case of AD, Berberine is capable of checking the activity of different enzymes required in pathogenesis [65]. Moreover, Berberine also capable of solubilizing amyloid aggregation in the AD mouse model [66]. Several cell-based studies demonstrated that Berberine is very useful to save neuronal cells from H_2O_2, glutamate, and cobalt chloride-induced neurotoxicity [65, 67, 68]. During cobalt chloride-induced hypoxic condition, berberine act as a free radical scavenger and prevent various apoptosis-promoting signalling molecule, consequently exhibiting neuroprotection [69]. Though, few authors have reported that Berberine has serious side effects, which involve deterioration in DA neurons because of 6-hydroxydopamine-induced cytotoxicity [70]. Neuroprotective effect of Berberine due to the activation of the PI3K/Akt/Nrf2 pathway by scavenging free radicals [71]. Also, Berberine displays anti-apoptotic results by diminishing the gene expression of caspases 1 and 3, bax, as well as upregulation of Bcl-2 [66]. Hsu *et al.* have revealed that berberine treatment inhibits hydrogen peroxide-induced neurotoxicity by decreasing the gene expression of p53, caspase, cyclin D1, and enhancing the expression of Bad (Bcl-2 associated agonist of cell death protein) [71]. In one study, It has been confirmed that berberine at nanomolar concentration favors cell survival and decreases OS by inhibiting the expression of various factors like cytochrome c, Bax (Bcl-2 associated X protein), and caspase. During ischemic stroke, Berberine treatment is beneficial that protect the neuronal cell from death by inhibiting potassium ions [72]. Moreover, Berberine removes the Htt protein aggregation through the autophagic mechanism [73, 74].

Curcumin

Curcumin is a principal active constituent of turmeric. Curcumin possesses several medicinal attributes, and consequently, it is used as a medication for metabolic disorders, immunological disorders, and hepatic disorders [75 - 77]. In the case of AD pathogenesis, curcumin can bind to aggregated amyloid through inhibiting NF-κB [75]. This was confirmed in an experiment in which about 214 compounds holding antioxidant characteristics were observed, and curcumin was affirmed to have a tremendous affinity towards amyloid plaques [78]. Curcumin administration in aged animals improves the cognitive impairments through an increment in CREB and BDNF (Brain-Derived Neurotropic Factor) levels [75]. Curcumin can clear amyloid plaque formation in both in-*vivo* and cell culture conditions [79]. In the PD model, curcumin promotes neurons regeneration through initiating Trk/PI3K signaling pathways, which elevate BDNF levels [80]. Curcumin is considered to function by diminishing TNF-α and caspase levels and concurrently raising BDNF levels [75, 81]. In one study, researcher-made curcumin nanoparticles mitigate the condition of cognitive impairment *via* recovering BDNF levels through Akt/GSK-3beta signaling pathways [82]. Curcumin is an encouraging secure, and inexpensive preventive measure for NDs treatment. According to Zhang *et al.*, it is clear that curcumin not only decreases the levels but also limits the maturation of amyloid-beta precursor protein (APP) in rodent neurons [83]. The underlying mechanism of clearance of plaque by curcumin is the inhibition of NF-κB, curcumin also capable of destabilizing the α-synuclein in the case of PD [81]. In one report, it has been seen that curcumin protects fruit fly from HD pathogenesis through inhibiting apoptosis.

Resveratrol

The source of resveratrol is grapes, peanuts, wine, and tea. Due to the presence of antioxidant and anti-inflammatory attributes in interest with NDs, resveratrol is known as a "miracle molecule" [84]. In various experimental models of cognitive defects, resveratrol treatment reversed the effect by reducing TNF-α and IL-1beta levels and raising BDNF levels in the hippocampus [84]. Moreover, resveratrol enhances IL-10 levels, which act as an anti-inflammatory by reducing TNF-α and NF-κB levels. Additionally, resveratrol also supports neuronal survival through activation of ERK1-2/CREB signaling [84, 85]. In the case of NDs, resveratrol gives neuroprotection by suppressing the glial cell activation. These cells are activated during NDs pathogenesis and started releasing inflammatory cytokines as well as different neurotoxic molecules, *e.g.*, nitric oxide, superoxide, *etc* [85]. In another study, it has been observed that resveratrol restore the cognitive function in the streptozotocin-induced model successfully [86]. Since the majority of the NDs are characterized by the presence of aggregated protein plaque in neurons, in such cases, resveratrol is beneficial for the destabilization of plaques [87]. In a study done a few years back, in which it was found that resveratrol inhibits 6-hydroxydopamine (6-OHDA) induced neurotoxicity through activation of Sirtuin 1 (SIRT-1) [88]. From the above report, it is clear that both resveratrol and SIRT-1 are essential to inhibit neuronal cell death. In this mechanism, resveratrol prevents the deacetylation of SIRT-1 substrates, *e.g.*, p53 and PGC-1α [89]. In one report, it has been observed that resveratrol exhibits neuroprotective effects by activation of the autophagy pathway (AMPK-SIRT1) in rotenone-induced in-*vitro* PD models [90].

Quercetin

This is a kind of flavonoid and present in several fruits and vegetables. It has been published that phytochemicals are used in complex diseases, *e.g.*, cancer, hepatic disorder, and NDs [91 - 93]. Quercetin (3,3′,4′,5,7-pentahydroxyflavone) has attracted the mind of researchers since the last decade because of its antioxidant, free radical scavenging attributes and is being extensively examined for antiproliferative capacity [94]. The antioxidant potential of quercetin has been explained due to the presence of the catechol group in the B ring and the OH group at position 3. This is very effective neuroprotective at the lowest concentration of 10 μM as well as at the highest of 30 μM. It has also been published that quercetin can cross the blood-brain barrier (BBB) and inhibit the H_2O_2 induced cytotoxicity [95]. Quercetin act as a kinase regulator in cells for the modulation of cellular function and gene expression [96]. On different cell lines, quercetin effects depend the cell type and treatment time and concentrations (higher than 100 μM) that may exhibit apoptosis, antiproliferative, cellular toxicity, and genotoxic activities. During neuroinflammation in NDs pathogenesis, quercetin treatments regulate NF-κB activity and show the anti-inflammatory effect [96].

Fisetin

Fisetin (3, 7, 3,4-tetrahydroxyflavone) is a flavonoid and active component of fruits and vegetables, *e.g.*, apples, grapes, kiwis, strawberries, onions, persimmons, and cucumbers [97, 98]. Various studies have proposed that fisetin is a biochemically and physiologically active phytochemical based phytomedicine, conferring several strong pharmacological potentials toward cancer, OS, inflammatory bowel disease, and NDs [99, 100]. In a recent report, it has been seen that fisetin protects adult mice from Aβ-induced neuroinflammation and neurodegeneration. In another study, fisetin exhibits antioxidant and neuroprotective effects by activating ERK in the ALS animal model [101].

Ginsenosides

Ginsenosides, the principal active ingredients of ginseng, are belong to the class of tetracyclic triterpene glycosides [102], and they have been used as a therapeutic and pharmacological outcome for the treatment of NDs [103 - 105]. Particularly, ginsenoside (GRg1) was confirmed that it could penetrate BBB and reach to the cortex, hippocampus, and striatum corpora. GRg1 provides neuroprotection by improving amyloid pathology, modulating the APP process, enhancing cognition, and activating PKA/CREB signaling [106]. Another ginsenoside compound, at 10 mg/kg for 7 days treatment, could decrease tau hyperphosphorylation and neurotoxicity, also improving the protein phosphatase 2A activity [107].

Sulforaphane (SFN)

Sulforaphane (SFN, 1-isothiocyanate-4-methylsulfinyl butane), a compound obtained from cruciferous vegetables [108]. SFN can easily be seen in blood plasma with peak concentration after 1 to 3 hours of treatments [109]. SFN has antioxidant potential because it is a potent inducer of the nuclear factor erythroid 2-related factor 2 (Nrf2)-antioxidant response element (ARE) pathway [110]. In mice neuroblastoma cell line (Neuro 2A cells), SFN protects cells from Aβ1-42 induced toxicity through the proteasome-dependent way [111]. SFN also acts as an anti-inflammatory through inhibition of NF-κB translocation in the macrophage cell line of the mouse [112, 113]. It also shows an anti-inflammatory effect by lowering the IFN-γ, MCP-1, and TNF-α, as well as enhancing IL-10 level [114].

Genistein

Genistein (4′,5,7-trihydroxyisoflavone) is commonly known as a phytoestrogen which belongs to the isoflavonoid class [115]. It obtains from soy and holds antioxidant and neuroprotective attributes [116]. Various pharmacokinetic investigations reported that genistein has poor oral bioavailability [117]. Interestingly, genistein is an agonist to the estrogen receptor (ER) β and has a structural resemblance with estrogen. ERβ is located in the brain associated with learning and memory [118, 119]. Moreover, genistein directly interacts with ERβ and show a neuroprotective effect also can enhance the memory deficit of NDs [120].

Allium and Allicin

It is an organosulfur compound, commonly present in high amounts in garlic as well as onions, and has neuroprotective properties [121]. Due to their antioxidant activities and allyl-containing sulfides group, they initiate stress-response pathways, consequently upregulates the neuroprotective proteins, *e.g.*, mitochondrial uncoupling proteins [122]. Furthermore, allicin acts on the plasma membrane of neurons and instigates transient receptor potential (TRP) ion channels in the neurons [47, 122]. Data shows that TRP channels perform a pivotal role in the regulation of OS and lysosome functions through controlling calcium ion influx and efflux [123]. Interestingly, TRP has a protective role in NDs; its overexpression prompts autophagy *via* initiating the mTOR pathway [124], consequently minimizing neuronal apoptosis.

Icariin

Icariin ($C_{33}H_{40}O_{15}$) is a flavonoid recognized as the principal bioactive phytochemical of *Epimedium brevicornum* Maxim, a classical Chinese herbal medicine [125, 126]. It has anti-inflammatory, antioxidant, and antidepressant [127, 128]. The bioavailability of Icariin is very poor because of the presence of prenyl-moiety [129]. Icariin has been observed to have various neuroprotective effects. It enhances the function and survival of neurons [130, 131] and instigates their self-renewal by neural stem cells [132]. The impacts of icariin in preventing Aβ deposition and Aβ induced neurotoxicity have been widely studied [133]; further, icariin also inhibits learning and memory deficit through increasing SOD activity and decreasing MDA levels [134].

Morin

Morin (3,5,7,2′,4′-pentahydroxyflavone) is a phytochemical and belongs to the flavonoid class obtained from plants of the Moraceae family. It is used as a herbal medicine in different countries. Earlier investigations have shown that morin displays antioxidant, anti-inflammatory, and antiproliferative results in both *in-vivo* and *in-vitro* experiments [135, 136]. Furthermore, morin inhibits the side-effects of OS by the instigation of the nuclear factor-erythroid related factor 2 (Nrf2) pathway [137]. In one report, it has been published that morin acts as a neuroprotective through multiple mechanisms in 1-methyl-4-phenyl-1,2-3,6-tetrahydropyridine (MPTP) induced mouse model of PD [138].

Baicalein

Baicalein (5,6,7-trihydroxyflavone) is a principal bioactive phytochemical and belongs to the flavone family constituent root of *Scutellaria baicalensis* Georgi [139]. It is Chinese traditional medicine; the root of *S. baicalensis* Georgi is used in the decoction for herbal medicine in the treatment of diseases related to the CNS. According to pharmacokinetic studies, it is confirmed that baicalein has a variety of pharmacological properties *e.g.*, antioxidant, anti-inflammatory, neuroprotective, stimulating neurogenesis, and differentiation action [140]. Based on these characteristics, baicalein exhibits therapeutic potential for AD and PD [141]. Both neurotrophic and neurogenesis effects of baicalein have been tested in a series of experiments for the investigation of the molecular mechanisms. Baicalein protects the hippocampus by radiation-induced damage through the modulation of OS and activation of neurotrophic factor-pCREB signaling [142]. Furthermore, baicalein also encourages neurogenesis in the MPTP model of PD through the activation of a series of genes such as SRY (sex-determining region Y)-box 8 (SOX8) and SRY (sex-determining region Y)-box 11 (SOX11) genes that can increase endogenous neurogenesis [141].

CONCLUSIONS

As we discussed various phytochemicals in this chapter and most of these are part of our daily life. The majority of the phytochemicals act as an antioxidant as well as anti-inflammatory. These phytochemicals control oxidative stress, neuroinflammation and provide neuroprotection during NDs pathogenesis (Fig. 1), as a consequence of these inhibitions that regulate glial activation, mitochondrial dysfunction, the release of inflammatory mediators, and ROS/RNS accumulation. But the question is that if we are using such kinds of phytochemicals routinely, why are we facing neurodegenerative disorders? The probable answer would be that we are not taking these phytochemicals in adequate quantity as well as the appropriate form. We don't have a standardized method for the use of such kinds of phytochemicals. The reason behind this ignorance is that we don't have enough clinical data regarding proper applications of phytochemicals.

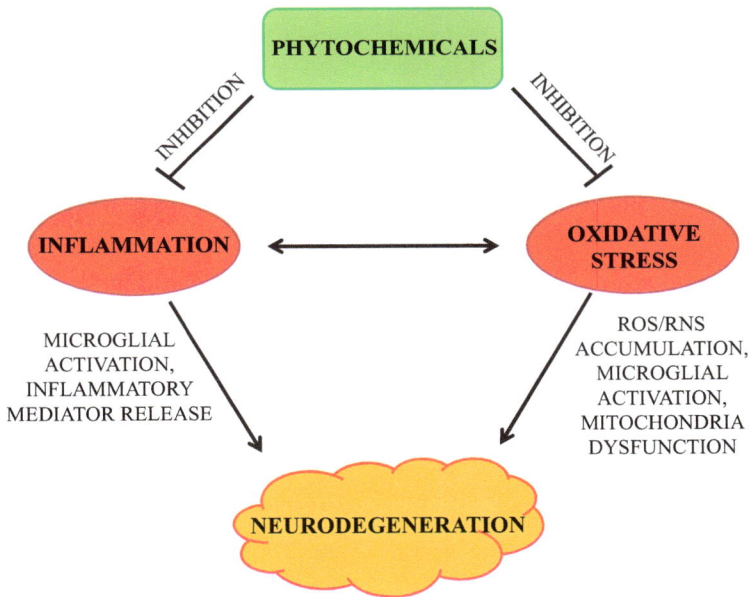

Fig. (1). Mode of action of phytochemicals.

FUTURE SUGGESTION

The phytochemicals mentioned in this chapter have been identified for their neuroprotective roles. Every polyphenol has distinct molecular targets; therefore, research is required regarding combinatorial therapies of these phytochemicals for the treatment of NDs. Moreover, The potential benefits of these phytochemicals should be well studied, as well as more extensive studies need to be conducted to establish the long-term effects and efficacy of using phytochemicals as therapeutics for neurodegenerative diseases.

CONSENT FOR PUBLICATION

Not applicable.

CONFLICT OF INTEREST

The authors declare no conflict of interest, financial or otherwise.

ACKNOWLEDGEMENTS

I would like to acknowledge Dr. D.S. Kothari Post-Doctoral Fellowship scheme of University Grant Commission (UGC), New Delhi, India for the award to Shambhu Sharan Tripathi (F.4-2/2006 (BSR)/BL/17-18/0381) and Prabhash Kumar Pandey (F.4-2/2006 (BSR)/BL/17-18/0442).

REFERENCES

[1]   Campbell IL, Krucker T, Steffensen S, *et al.* Structural and functional neuropathology in transgenic mice with CNS expression of IFN-alpha. Brain Res 1999; 835(1): 46-61.
[http://dx.doi.org/10.1016/S0006-8993(99)01328-1] [PMID: 10448195]

[2]   Nakamura T, Cho DH, Lipton SA. Redox regulation of protein misfolding, mitochondrial dysfunction, synaptic damage, and cell death in neurodegenerative diseases. Exp Neurol 2012; 238(1): 12-21.
[http://dx.doi.org/10.1016/j.expneurol.2012.06.032] [PMID: 22771760]

[3]   Falsone A, Falsone SF. Legal but lethal: functional protein aggregation at the verge of toxicity. Front Cell Neurosci 2015; 9: 45.
[http://dx.doi.org/10.3389/fncel.2015.00045] [PMID: 25741240]

[4]   Zhou L, Miranda-Saksena M, Saksena NK. Viruses and neurodegeneration. Virol J 2013; 10: 172.
[http://dx.doi.org/10.1186/1743-422X-10-172] [PMID: 23724961]

[5]   Popescu B F, Pirko I, Lucchinetti C F. Pathology of multiple sclerosis: where do we stand? Continuum (Minneap Minn) 2013; 19(4 Multiple Sclerosis): 901-21.
[http://dx.doi.org/10.1212/01.CON.0000433291.23091.65]

[6]   Mayo L, Quintana FJ, Weiner HL. The innate immune system in demyelinating disease. Immunol Rev 2012; 248(1): 170-87.
[http://dx.doi.org/10.1111/j.1600-065X.2012.01135.x] [PMID: 22725961]

[7]   Wyss-Coray T. Ageing, neurodegeneration and brain rejuvenation. Nature 2016; 539(7628): 180-6.
[http://dx.doi.org/10.1038/nature20411] [PMID: 27830812]

[8]   Tuite MF, Melki R. Protein misfolding and aggregation in ageing and disease: molecular processes and therapeutic perspectives. Prion 2007; 1(2): 116-20.
[http://dx.doi.org/10.4161/pri.1.2.4651] [PMID: 19164925]

[9]   Sweeney P, Park H, Baumann M, *et al.* Protein misfolding in neurodegenerative diseases: implications and strategies. Transl Neurodegener 2017; 6: 6.
[http://dx.doi.org/10.1186/s40035-017-0077-5] [PMID: 28293421]

[10]   Davis AA, Leyns CEG, Holtzman DM. Intercellular Spread of Protein Aggregates in

Neurodegenerative Disease. Annu Rev Cell Dev Biol 2018; 34: 545-68.
[http://dx.doi.org/10.1146/annurev-cellbio-100617-062636] [PMID: 30044648]

[11]   Qiu C, Kivipelto M, von Strauss E. Epidemiology of Alzheimer's disease: occurrence, determinants, and strategies toward intervention. Dialogues Clin Neurosci 2009; 11(2): 111-28.
[http://dx.doi.org/10.31887/DCNS.2009.11.2/cqiu] [PMID: 19585947]

[12]   Niu H, Álvarez-Álvarez I, Guillén-Grima F, Aguinaga-Ontoso I. Prevalence and incidence of Alzheimer's disease in Europe: A meta-analysis. Neurologia 2017; 32(8): 523-32.
[http://dx.doi.org/10.1016/j.nrl.2016.02.016] [PMID: 27130306]

[13]   Rizek P, Kumar N, Jog MS. An update on the diagnosis and treatment of Parkinson disease. CMAJ 2016; 188(16): 1157-65.
[http://dx.doi.org/10.1503/cmaj.151179] [PMID: 27221269]

[14]   DeMaagd G, Philip A. Parkinson's Disease and Its Management: Part 1: Disease Entity, Risk Factors, Pathophysiology, Clinical Presentation, and Diagnosis. P&T 2015; 40(8): 504-32.
[PMID: 26236139]

[15]   Harper PS. The epidemiology of Huntington's disease. Hum Genet 1992; 89(4): 365-76.
[http://dx.doi.org/10.1007/BF00194305] [PMID: 1535611]

[16]   Talbott EO, Malek AM, Lacomis D. The epidemiology of amyotrophic lateral sclerosis. Handb Clin Neurol 2016; 138: 225-38.
[http://dx.doi.org/10.1016/B978-0-12-802973-2.00013-6] [PMID: 27637961]

[17]   Sheng B, Wang X, Su B, *et al.* Impaired mitochondrial biogenesis contributes to mitochondrial dysfunction in Alzheimer's disease. J Neurochem 2012; 120(3): 419-29.
[http://dx.doi.org/10.1111/j.1471-4159.2011.07581.x] [PMID: 22077634]

[18]   Hattingen E, Magerkurth J, Pilatus U, *et al.* Phosphorus and proton magnetic resonance spectroscopy demonstrates mitochondrial dysfunction in early and advanced Parkinson's disease. Brain 2009; 132(Pt 12): 3285-97.
[http://dx.doi.org/10.1093/brain/awp293] [PMID: 19952056]

[19]   Reddy PH, Shirendeb UP. Mutant huntingtin, abnormal mitochondrial dynamics, defective axonal transport of mitochondria, and selective synaptic degeneration in Huntington's disease. Biochim Biophys Acta 2012; 1822(2): 101-10.
[http://dx.doi.org/10.1016/j.bbadis.2011.10.016] [PMID: 22080977]

[20]   Cozzolino M, Carrì MT. Mitochondrial dysfunction in ALS. Prog Neurobiol 2012; 97(2): 54-66.
[http://dx.doi.org/10.1016/j.pneurobio.2011.06.003] [PMID: 21827820]

[21]   Mancuso M, Coppede F, Migliore L, Siciliano G, Murri L. Mitochondrial dysfunction, oxidative stress and neurodegeneration. J Alzheimers Dis 2006; 10(1): 59-73.
[http://dx.doi.org/10.3233/JAD-2006-10110] [PMID: 16988483]

[22]   Guo C, Sun L, Chen X, Zhang D. Oxidative stress, mitochondrial damage and neurodegenerative diseases. Neural Regen Res 2013; 8(21): 2003-14.
[PMID: 25206509]

[23]   Elfawy HA, Das B. Crosstalk between mitochondrial dysfunction, oxidative stress, and age related neurodegenerative disease: Etiologies and therapeutic strategies. Life Sci 2019; 218: 165-84.
[http://dx.doi.org/10.1016/j.lfs.2018.12.029] [PMID: 30578866]

[24]   Chen X, Guo C, Kong J. Oxidative stress in neurodegenerative diseases. Neural Regen Res 2012; 7(5): 376-85.
[PMID: 25774178]

[25]   Uttara B, Singh AV, Zamboni P, Mahajan RT. Oxidative stress and neurodegenerative diseases: a review of upstream and downstream antioxidant therapeutic options. Curr Neuropharmacol 2009; 7(1): 65-74.
[http://dx.doi.org/10.2174/157015909787602823] [PMID: 19721819]

[26]   Singh A, Kukreti R, Saso L, Kukreti S. Oxidative Stress: A Key Modulator in Neurodegenerative Diseases. Molecules 2019; 24(8): E1583.
[http://dx.doi.org/10.3390/molecules24081583] [PMID: 31013638]

[27]   Praticò D, Uryu K, Leight S, Trojanoswki JQ, Lee VM. Increased lipid peroxidation precedes amyloid plaque formation in an animal model of Alzheimer amyloidosis. J Neurosci 2001; 21(12): 4183-7.
[http://dx.doi.org/10.1523/JNEUROSCI.21-12-04183.2001] [PMID: 11404403]

[28]   Shichiri M. The role of lipid peroxidation in neurological disorders. J Clin Biochem Nutr 2014; 54(3): 151-60.
[http://dx.doi.org/10.3164/jcbn.14-10] [PMID: 24895477]

[29]   Gonos ES, Kapetanou M, Sereikaite J, et al. Origin and pathophysiology of protein carbonylation, nitration and chlorination in age-related brain diseases and aging. Aging (Albany NY) 2018; 10(5): 868-901.
[http://dx.doi.org/10.18632/aging.101450] [PMID: 29779015]

[30]   Therond P. [Oxidative stress and damages to biomolecules (lipids, proteins, DNA)]. Ann Pharm Fr 2006; 64(6): 383-9. [Oxidative stress and damages to biomolecules (lipids, proteins, DNA)].
[http://dx.doi.org/10.1016/S0003-4509(06)75333-0] [PMID: 17119467]

[31]   Li R, Jia Z, Trush MA. Defining ROS in Biology and Medicine. React Oxyg Species (Apex) 2016; 1(1): 9-21.
[http://dx.doi.org/10.20455/ros.2016.803] [PMID: 29707643]

[32]   Wang X, Wang W, Li L, Perry G, Lee HG, Zhu X. Oxidative stress and mitochondrial dysfunction in Alzheimer's disease. Biochim Biophys Acta 2014; 1842(8): 1240-7.
[http://dx.doi.org/10.1016/j.bbadis.2013.10.015] [PMID: 24189435]

[33]   Sultana R, Perluigi M, Newman SF, et al. Redox proteomic analysis of carbonylated brain proteins in mild cognitive impairment and early Alzheimer's disease. Antioxid Redox Signal 2010; 12(3): 327-36.
[http://dx.doi.org/10.1089/ars.2009.2810] [PMID: 19686046]

[34]   Gabbita SP, Lovell MA, Markesbery WR. Increased nuclear DNA oxidation in the brain in Alzheimer's disease. J Neurochem 1998; 71(5): 2034-40.
[http://dx.doi.org/10.1046/j.1471-4159.1998.71052034.x] [PMID: 9798928]

[35]   Dias V, Junn E, Mouradian MM. The role of oxidative stress in Parkinson's disease. J Parkinsons Dis 2013; 3(4): 461-91.
[http://dx.doi.org/10.3233/JPD-130230] [PMID: 24252804]

[36]   Zhou C, Huang Y, Przedborski S. Oxidative stress in Parkinson's disease: a mechanism of pathogenic and therapeutic significance. Ann N Y Acad Sci 2008; 1147: 93-104.
[http://dx.doi.org/10.1196/annals.1427.023] [PMID: 19076434]

[37]   Alam ZI, Daniel SE, Lees AJ, Marsden DC, Jenner P, Halliwell B. A generalised increase in protein carbonyls in the brain in Parkinson's but not incidental Lewy body disease. J Neurochem 1997; 69(3): 1326-9.
[http://dx.doi.org/10.1046/j.1471-4159.1997.69031326.x] [PMID: 9282961]

[38]   Floor E, Wetzel MG. Increased protein oxidation in human substantia nigra pars compacta in comparison with basal ganglia and prefrontal cortex measured with an improved dinitrophenylhydrazine assay. J Neurochem 1998; 70(1): 268-75.
[http://dx.doi.org/10.1046/j.1471-4159.1998.70010268.x] [PMID: 9422371]

[39]   Good PF, Hsu A, Werner P, Perl DP, Olanow CW. Protein nitration in Parkinson's disease. J Neuropathol Exp Neurol 1998; 57(4): 338-42.
[http://dx.doi.org/10.1097/00005072-199804000-00006] [PMID: 9600227]

[40]   Giasson BI, Duda JE, Murray IV, et al. Oxidative damage linked to neurodegeneration by selective alpha-synuclein nitration in syncleinopathy lesions. Science 2000; 290(5493): 985-9.
[http://dx.doi.org/10.1126/science.290.5493.985] [PMID: 11062131]

[41]   Tan EK, Srivastava AK, Arnold WD, Singh MP, Zhang Y. Neurodegeneration: etiologies and new therapies. BioMed Res Int 2015; 2015: 272630.
[http://dx.doi.org/10.1155/2015/272630] [PMID: 25685777]

[42]   Richards RI, Robertson SA, Kastner DL. Neurodegenerative diseases have genetic hallmarks of autoinflammatory disease. Hum Mol Genet 2018; 27(R2): R108-18.
[http://dx.doi.org/10.1093/hmg/ddy139] [PMID: 29684205]

[43]   Zhang YJ, Gan RY, Li S, *et al.* Antioxidant Phytochemicals for the Prevention and Treatment of Chronic Diseases. Molecules 2015; 20(12): 21138-56.
[http://dx.doi.org/10.3390/molecules201219753] [PMID: 26633317]

[44]   Islam MA, Alam F, Solayman M, Khalil MI, Kamal MA, Gan SH. Dietary Phytochemicals: Natural Swords Combating Inflammation and Oxidation-Mediated Degenerative Diseases. Oxid Med Cell Longev 2016; 2016: 5137431.
[http://dx.doi.org/10.1155/2016/5137431] [PMID: 27721914]

[45]   Köhle C, Bock KW. Activation of coupled Ah receptor and Nrf2 gene batteries by dietary phytochemicals in relation to chemoprevention. Biochem Pharmacol 2006; 72(7): 795-805.
[http://dx.doi.org/10.1016/j.bcp.2006.04.017] [PMID: 16780804]

[46]   Dietrich C. Antioxidant Functions of the Aryl Hydrocarbon Receptor. Stem Cells Int 2016; 2016: 7943495.
[http://dx.doi.org/10.1155/2016/7943495] [PMID: 27829840]

[47]   Kumar GP, Khanum F. Neuroprotective potential of phytochemicals. Pharmacogn Rev 2012; 6(12): 81-90.
[http://dx.doi.org/10.4103/0973-7847.99898] [PMID: 23055633]

[48]   Velmurugan BK, Rathinasamy B, Lohanathan BP, Thiyagarajan V, Weng CF. Neuroprotective Role of Phytochemicals. Molecules 2018; 23(10): E2485.
[http://dx.doi.org/10.3390/molecules23102485] [PMID: 30262792]

[49]   Bastianetto S, Zheng WH, Quirion R. Neuroprotective abilities of resveratrol and other red wine constituents against nitric oxide-related toxicity in cultured hippocampal neurons. Br J Pharmacol 2000; 131(4): 711-20.
[http://dx.doi.org/10.1038/sj.bjp.0703626] [PMID: 11030720]

[50]   Singh NA, Mandal AK, Khan ZA. Potential neuroprotective properties of epigallocatechin-3-gallate (EGCG). Nutr J 2016; 15(1): 60.
[http://dx.doi.org/10.1186/s12937-016-0179-4] [PMID: 27268025]

[51]   Guo S, Bezard E, Zhao B. Protective effect of green tea polyphenols on the SH-SY5Y cells against 6-OHDA induced apoptosis through ROS-NO pathway. Free Radic Biol Med 2005; 39(5): 682-95.
[http://dx.doi.org/10.1016/j.freeradbiomed.2005.04.022] [PMID: 16085186]

[52]   Weinreb O, Mandel S, Amit T, Youdim MB. Neurological mechanisms of green tea polyphenols in Alzheimer's and Parkinson's diseases. J Nutr Biochem 2004; 15(9): 506-16.
[http://dx.doi.org/10.1016/j.jnutbio.2004.05.002] [PMID: 15350981]

[53]   Lim HJ, Shim SB, Jee SW, *et al.* Green tea catechin leads to global improvement among Alzheimer's disease-related phenotypes in NSE/hAPP-C105 Tg mice. J Nutr Biochem 2013; 24(7): 1302-13.
[http://dx.doi.org/10.1016/j.jnutbio.2012.10.005] [PMID: 23333093]

[54]   Hu G, Bidel S, Jousilahti P, Antikainen R, Tuomilehto J. Coffee and tea consumption and the risk of Parkinson's disease. Mov Disord 2007; 22(15): 2242-8.
[http://dx.doi.org/10.1002/mds.21706] [PMID: 17712848]

[55]   Choi JY, Park CS, Kim DJ, *et al.* Prevention of nitric oxide-mediated 1-methyl-4-phenyl-1,2-3,6-tetrahydropyridine-induced Parkinson's disease in mice by tea phenolic epigallocatechin 3-gallate. Neurotoxicology 2002; 23(3): 367-74.
[http://dx.doi.org/10.1016/S0161-813X(02)00079-7] [PMID: 12387363]

[56] Zhou T, Zhu M, Liang Z. (-)-Epigallocatechin-3-gallate modulates peripheral immunity in the MPTP-induced mouse model of Parkinson's disease. Mol Med Rep 2018; 17(4): 4883-8.
[http://dx.doi.org/10.3892/mmr.2018.8470] [PMID: 29363729]

[57] Koh SH, Kim SH, Kwon H, *et al.* Epigallocatechin gallate protects nerve growth factor differentiated PC12 cells from oxidative-radical-stress-induced apoptosis through its effect on phosphoinositide 3-kinase/Akt and glycogen synthase kinase-3. Brain Res Mol Brain Res 2003; 118(1-2): 72-81.
[http://dx.doi.org/10.1016/j.molbrainres.2003.07.003] [PMID: 14559356]

[58] Liu M, Chen F, Sha L, *et al.* (-)-Epigallocatechin-3-gallate ameliorates learning and memory deficits by adjusting the balance of TrkA/p75NTR signaling in APP/PS1 transgenic mice. Mol Neurobiol 2014; 49(3): 1350-63.
[http://dx.doi.org/10.1007/s12035-013-8608-2] [PMID: 24356899]

[59] Kalaiselvi P, Rajashree K, Bharathi Priya L, Padma VV. Cytoprotective effect of epigallocatechin--gallate against deoxynivalenol-induced toxicity through anti-oxidative and anti-inflammatory mechanisms in HT-29 cells. Food Chem Toxicol 2013; 56: 110-8.
[http://dx.doi.org/10.1016/j.fct.2013.01.042] [PMID: 23410590]

[60] Han J, Wang M, Jing X, Shi H, Ren M, Lou H. (-)-Epigallocatechin gallate protects against cerebral ischemia-induced oxidative stress *via* Nrf2/ARE signaling. Neurochem Res 2014; 39(7): 1292-9.
[http://dx.doi.org/10.1007/s11064-014-1311-5] [PMID: 24792731]

[61] Ehrnhoefer DE, Duennwald M, Markovic P, *et al.* Green tea (-)-epigallocatechin-gallate modulates early events in huntingtin misfolding and reduces toxicity in Huntington's disease models. Hum Mol Genet 2006; 15(18): 2743-51.
[http://dx.doi.org/10.1093/hmg/ddl210] [PMID: 16893904]

[62] Andrich K, Bieschke J. The Effect of. The Effect of (-)-Epigallo-catechin-(3)-gallate on Amyloidogenic Proteins Suggests a Common Mechanism. Adv Exp Med Biol 2015; 863: 139-61.
[http://dx.doi.org/10.1007/978-3-319-18365-7_7] [PMID: 26092630]

[63] Ji HF, Shen L. Berberine: a potential multipotent natural product to combat Alzheimer's disease. Molecules 2011; 16(8): 6732-40.
[http://dx.doi.org/10.3390/molecules16086732] [PMID: 21829148]

[64] Durairajan SS, Liu LF, Lu JH, *et al.* Berberine ameliorates β-amyloid pathology, gliosis, and cognitive impairment in an Alzheimer's disease transgenic mouse model. Neurobiol Aging 2012; 33(12): 2903-19.
[http://dx.doi.org/10.1016/j.neurobiolaging.2012.02.016] [PMID: 22459600]

[65] Hsu YY, Tseng YT, Lo YC. Berberine, a natural antidiabetes drug, attenuates glucose neurotoxicity and promotes Nrf2-related neurite outgrowth. Toxicol Appl Pharmacol 2013; 272(3): 787-96.
[http://dx.doi.org/10.1016/j.taap.2013.08.008] [PMID: 23954465]

[66] Asai M, Iwata N, Yoshikawa A, *et al.* Berberine alters the processing of Alzheimer's amyloid precursor protein to decrease Abeta secretion. Biochem Biophys Res Commun 2007; 352(2): 498-502.
[http://dx.doi.org/10.1016/j.bbrc.2006.11.043] [PMID: 17125739]

[67] Simões Pires EN, Frozza RL, Hoppe JB, Menezes BdeM, Salbego CG. Berberine was neuroprotective against an in *vitro* model of brain ischemia: survival and apoptosis pathways involved. Brain Res 2014; 1557: 26-33.
[http://dx.doi.org/10.1016/j.brainres.2014.02.021] [PMID: 24560603]

[68] Cui HS, Matsumoto K, Murakami Y, Hori H, Zhao Q, Obi R. Berberine exerts neuroprotective actions against in *vitro* ischemia-induced neuronal cell damage in organotypic hippocampal slice cultures: involvement of B-cell lymphoma 2 phosphorylation suppression. Biol Pharm Bull 2009; 32(1): 79-85.
[http://dx.doi.org/10.1248/bpb.32.79] [PMID: 19122285]

[69] Zhang Q, Qian Z, Pan L, Li H, Zhu H. Hypoxia-inducible factor 1 mediates the anti-apoptosis of berberine in neurons during hypoxia/ischemia. Acta Physiol Hung 2012; 99(3): 311-23.

[http://dx.doi.org/10.1556/APhysiol.99.2012.3.8] [PMID: 22982719]

[70]   Kwon IH, Choi HS, Shin KS, *et al.* Effects of berberine on 6-hydroxydopamine-induced neurotoxicity in PC12 cells and a rat model of Parkinson's disease. Neurosci Lett 2010; 486(1): 29-33.
[http://dx.doi.org/10.1016/j.neulet.2010.09.038] [PMID: 20851167]

[71]   Hsu YY, Chen CS, Wu SN, Jong YJ, Lo YC. Berberine activates Nrf2 nuclear translocation and protects against oxidative damage *via* a phosphatidylinositol 3-kinase/Akt-dependent mechanism in NSC34 motor neuron-like cells. Eur J Pharm Sci 2012; 46(5): 415-25.
[http://dx.doi.org/10.1016/j.ejps.2012.03.004] [PMID: 22469516]

[72]   Maleki SN, Aboutaleb N, Souri F. Berberine confers neuroprotection in coping with focal cerebral ischemia by targeting inflammatory cytokines. J Chem Neuroanat 2018; 87: 54-9.
[http://dx.doi.org/10.1016/j.jchemneu.2017.04.008] [PMID: 28495517]

[73]   Jiang W, Wei W, Gaertig MA, Li S, Li XJ. Therapeutic Effect of Berberine on Huntington's Disease Transgenic Mouse Model. PLoS One 2015; 10(7): e0134142.
[http://dx.doi.org/10.1371/journal.pone.0134142] [PMID: 26225560]

[74]   Fan D, Liu L, Wu Z, Cao M. Combating Neurodegenerative Diseases with the Plant Alkaloid Berberine: Molecular Mechanisms and Therapeutic Potential. Curr Neuropharmacol 2019; 17(6): 563-79.
[http://dx.doi.org/10.2174/1570159X16666180419141613] [PMID: 29676231]

[75]   Nam SM, Choi JH, Yoo DY, *et al.* Effects of curcumin (Curcuma longa) on learning and spatial memory as well as cell proliferation and neuroblast differentiation in adult and aged mice by upregulating brain-derived neurotrophic factor and CREB signaling. J Med Food 2014; 17(6): 641-9.
[http://dx.doi.org/10.1089/jmf.2013.2965] [PMID: 24712702]

[76]   Soleimani H, Amini A, Taheri S, *et al.* The effect of combined photobiomodulation and curcumin on skin wound healing in type I diabetes in rats. J Photochem Photobiol B 2018; 181: 23-30.
[http://dx.doi.org/10.1016/j.jphotobiol.2018.02.023] [PMID: 29486459]

[77]   El-Bahr SM. Effect of curcumin on hepatic antioxidant enzymes activities and gene expressions in rats intoxicated with aflatoxin B1. Phytother Res 2015; 29(1): 134-40.
[http://dx.doi.org/10.1002/ptr.5239] [PMID: 25639897]

[78]   Kim H, Park BS, Lee KG, *et al.* Effects of naturally occurring compounds on fibril formation and oxidative stress of beta-amyloid. J Agric Food Chem 2005; 53(22): 8537-41.
[http://dx.doi.org/10.1021/jf051985c] [PMID: 16248550]

[79]   Wang JQ, Xiong TT, Zhou J, *et al.* Enzymatic formation of curcumin in *vitro* and in *vivo.* Nano Res 2018; 11: 3453-61.
[http://dx.doi.org/10.1007/s12274-018-1994-z]

[80]   Yang J, Song S, Li J, Liang T. Neuroprotective effect of curcumin on hippocampal injury in 6-OHD--induced Parkinson's disease rat. Pathol Res Pract 2014; 210(6): 357-62.
[http://dx.doi.org/10.1016/j.prp.2014.02.005] [PMID: 24642369]

[81]   Liu D, Wang Z, Gao Z, *et al.* Effects of curcumin on learning and memory deficits, BDNF, and ERK protein expression in rats exposed to chronic unpredictable stress. Behav Brain Res 2014; 271: 116-21.
[http://dx.doi.org/10.1016/j.bbr.2014.05.068] [PMID: 24914461]

[82]   Hoppe JB, Coradini K, Frozza RL, *et al.* Free and nanoencapsulated curcumin suppress β-amyloi--induced cognitive impairments in rats: involvement of BDNF and Akt/GSK-3β signaling pathway. Neurobiol Learn Mem 2013; 106: 134-44.
[http://dx.doi.org/10.1016/j.nlm.2013.08.001] [PMID: 23954730]

[83]   Zhang C, Browne A, Child D, Tanzi RE. Curcumin decreases amyloid-beta peptide levels by attenuating the maturation of amyloid-beta precursor protein. J Biol Chem 2010; 285(37): 28472-80.
[http://dx.doi.org/10.1074/jbc.M110.133520] [PMID: 20622013]

[84]   Anastácio JR, Netto CA, Castro CC, *et al.* Resveratrol treatment has neuroprotective effects and

prevents cognitive impairment after chronic cerebral hypoperfusion. Neurol Res 2014; 36(7): 627-33.
[http://dx.doi.org/10.1179/1743132813Y.0000000293] [PMID: 24620966]

[85]   Ma T, Tan MS, Yu JT, Tan L. Resveratrol as a therapeutic agent for Alzheimer's disease. BioMed Res
         Int 2014; 2014: 350516.
         [http://dx.doi.org/10.1155/2014/350516] [PMID: 25525597]

[86]   Tian Z, Wang J, Xu M, Wang Y, Zhang M, Zhou Y. Resveratrol Improves Cognitive Impairment by
         Regulating Apoptosis and Synaptic Plasticity in Streptozotocin-Induced Diabetic Rats. Cell Physiol
         Biochem 2016; 40(6): 1670-7.
         [http://dx.doi.org/10.1159/000453216] [PMID: 28006780]

[87]   Karuppagounder SS, Pinto JT, Xu H, Chen HL, Beal MF, Gibson GE. Dietary supplementation with
         resveratrol reduces plaque pathology in a transgenic model of Alzheimer's disease. Neurochem Int
         2009; 54(2): 111-8.
         [http://dx.doi.org/10.1016/j.neuint.2008.10.008] [PMID: 19041676]

[88]   Albani D, Polito L, Batelli S, et al. The SIRT1 activator resveratrol protects SK-N-BE cells from
         oxidative stress and against toxicity caused by alpha-synuclein or amyloid-beta (1-42) peptide. J
         Neurochem 2009; 110(5): 1445-56.
         [http://dx.doi.org/10.1111/j.1471-4159.2009.06228.x] [PMID: 19558452]

[89]   Yazir Y, Utkan T, Gacar N, Aricioglu F. Resveratrol exerts anti-inflammatory and neuroprotective
         effects to prevent memory deficits in rats exposed to chronic unpredictable mild stress. Physiol Behav
         2015; 138: 297-304.
         [http://dx.doi.org/10.1016/j.physbeh.2014.10.010] [PMID: 25455865]

[90]   Wu Y, Li X, Zhu JX, et al. Resveratrol-activated AMPK/SIRT1/autophagy in cellular models of
         Parkinson's disease. Neurosignals 2011; 19(3): 163-74.
         [http://dx.doi.org/10.1159/000328516] [PMID: 21778691]

[91]   Tao SF, He HF, Chen Q. Quercetin inhibits proliferation and invasion acts by up-regulating miR-146a
         in human breast cancer cells. Mol Cell Biochem 2015; 402(1-2): 93-100.
         [http://dx.doi.org/10.1007/s11010-014-2317-7] [PMID: 25596948]

[92]   Godoy JA, Lindsay CB, Quintanilla RA, Carvajal FJ, Cerpa W, Inestrosa NC. Quercetin Exerts
         Differential Neuroprotective Effects Against H_2O_2 and Aβ Aggregates in Hippocampal Neurons: the
         Role of Mitochondria. Mol Neurobiol 2017; 54(9): 7116-28.
         [http://dx.doi.org/10.1007/s12035-016-0203-x] [PMID: 27796749]

[93]   Padma VV, Baskaran R, Roopesh RS, Poornima P. Quercetin attenuates lindane induced oxidative
         stress in Wistar rats. Mol Biol Rep 2012; 39(6): 6895-905.
         [http://dx.doi.org/10.1007/s11033-012-1516-0] [PMID: 22302394]

[94]   Lesjak M, Beara I, Simin N, et al. N., M.-D., Lesjak M., Beara I., Simin N., Pintac D., Majkic T.,
         Bekvalac K., Orcic D., Mimica-Dukic N. J Funct Foods 2018; 40: 68-5.
         [http://dx.doi.org/10.1016/j.jff.2017.10.047]

[95]   Heo HJ, Lee CY. Protective effects of quercetin and vitamin C against oxidative stress-induced
         neurodegeneration. J Agric Food Chem 2004; 52(25): 7514-7.
         [http://dx.doi.org/10.1021/jf049243r] [PMID: 15675797]

[96]   Dajas F, Abin-Carriquiry JA, Arredondo F, et al. Quercetin in brain diseases: Potential and limits.
         Neurochem Int 2015; 89: 140-8.
         [http://dx.doi.org/10.1016/j.neuint.2015.07.002] [PMID: 26160469]

[97]   Khan N, Syed DN, Ahmad N, Mukhtar H. Fisetin: a dietary antioxidant for health promotion. Antioxid
         Redox Signal 2013; 19(2): 151-62.
         [http://dx.doi.org/10.1089/ars.2012.4901] [PMID: 23121441]

[98]   Arai Y, Watanabe S, Kimira M, Shimoi K, Mochizuki R, Kinae N. Dietary intakes of flavonols,
         flavones and isoflavones by Japanese women and the inverse correlation between quercetin intake and

plasma LDL cholesterol concentration. J Nutr 2000; 130(9): 2243-50.
[http://dx.doi.org/10.1093/jn/130.9.2243] [PMID: 10958819]

[99] Aruoma OI. Methodological considerations for characterizing potential antioxidant actions of bioactive components in plant foods. Mutat Res 2003; 523-524: 9-20.
[http://dx.doi.org/10.1016/S0027-5107(02)00317-2] [PMID: 12628499]

[100] Marković ZS, Mentus SV, Dimitrić Marković JM. Electrochemical and density functional theory study on the reactivity of fisetin and its radicals: implications on in *vitro* antioxidant activity. J Phys Chem A 2009; 113(51): 14170-9.
[http://dx.doi.org/10.1021/jp907071v] [PMID: 19954196]

[101] Wang TH, Wang SY, Wang XD, *et al.* Fisetin Exerts Antioxidant and Neuroprotective Effects in Multiple Mutant hSOD1 Models of Amyotrophic Lateral Sclerosis by Activating ERK. Neuroscience 2018; 379: 152-66.
[http://dx.doi.org/10.1016/j.neuroscience.2018.03.008] [PMID: 29559385]

[102] Razgonova MP, Veselov VV, Zakharenko AM, *et al.* Panax ginseng components and the pathogenesis of Alzheimer's disease (Review). Mol Med Rep 2019; 19(4): 2975-98. [Review].
[http://dx.doi.org/10.3892/mmr.2019.9972] [PMID: 30816465]

[103] Kim HJ, Kim P, Shin CY. A comprehensive review of the therapeutic and pharmacological effects of ginseng and ginsenosides in central nervous system. J Ginseng Res 2013; 37(1): 8-29.
[http://dx.doi.org/10.5142/jgr.2013.37.8] [PMID: 23717153]

[104] Rajabian A, Rameshrad M, Hosseinzadeh H. Therapeutic potential of Panax ginseng and its constituents, ginsenosides and gintonin, in neurological and neurodegenerative disorders: a patent review. Expert Opin Ther Pat 2019; 29(1): 55-72.
[http://dx.doi.org/10.1080/13543776.2019.1556258] [PMID: 30513224]

[105] Zheng M, Xin Y, Li Y, *et al.* Ginsenosides: A Potential Neuroprotective Agent. BioMed Res Int 2018; 2018: 8174345.
[http://dx.doi.org/10.1155/2018/8174345] [PMID: 29854792]

[106] Fang F, Chen X, Huang T, Lue LF, Luddy JS, Yan SS. Multi-faced neuroprotective effects of Ginsenoside Rg1 in an Alzheimer mouse model. Biochim Biophys Acta 2012; 1822(2): 286-92.
[http://dx.doi.org/10.1016/j.bbadis.2011.10.004] [PMID: 22015470]

[107] Li L, Liu J, Yan X, *et al.* Protective effects of ginsenoside Rd against okadaic acid-induced neurotoxicity in *vivo* and in *vitro*. J Ethnopharmacol 2011; 138(1): 135-41.
[http://dx.doi.org/10.1016/j.jep.2011.08.068] [PMID: 21945003]

[108] McNaughton SA, Marks GC. Development of a food composition database for the estimation of dietary intakes of glucosinolates, the biologically active constituents of cruciferous vegetables. Br J Nutr 2003; 90(3): 687-97.
[http://dx.doi.org/10.1079/BJN2003917] [PMID: 13129476]

[109] Egner PA, Chen JG, Wang JB, *et al.* Bioavailability of Sulforaphane from two broccoli sprout beverages: results of a short-term, cross-over clinical trial in Qidong, China. Cancer Prev Res (Phila) 2011; 4(3): 384-95.
[http://dx.doi.org/10.1158/1940-6207.CAPR-10-0296] [PMID: 21372038]

[110] Sun Y, Yang T, Mao L, Zhang F. Sulforaphane Protects against Brain Diseases: Roles of Cytoprotective Enzymes. Austin J Cerebrovasc Dis Stroke 2017; 4(1): 1054.
[PMID: 29619410]

[111] Park HM, Kim JA, Kwak MK. Protection against amyloid beta cytotoxicity by sulforaphane: role of the proteasome. Arch Pharm Res 2009; 32(1): 109-15.
[http://dx.doi.org/10.1007/s12272-009-1124-2] [PMID: 19183883]

[112] Heiss E, Gerhäuser C. Time-dependent modulation of thioredoxin reductase activity might contribute to sulforaphane-mediated inhibition of NF-kappaB binding to DNA. Antioxid Redox Signal 2005;

7(11-12): 1601-11.
[http://dx.doi.org/10.1089/ars.2005.7.1601] [PMID: 16356123]

[113] Heiss E, Herhaus C, Klimo K, Bartsch H, Gerhäuser C. Nuclear factor kappa B is a molecular target for sulforaphane-mediated anti-inflammatory mechanisms. J Biol Chem 2001; 276(34): 32008-15.
[http://dx.doi.org/10.1074/jbc.M104794200] [PMID: 11410599]

[114] Holloway PM, Gillespie S, Becker F, *et al.* Sulforaphane induces neurovascular protection against a systemic inflammatory challenge *via* both Nrf2-dependent and independent pathways. Vascul Pharmacol 2016; 85: 29-38.
[http://dx.doi.org/10.1016/j.vph.2016.07.004] [PMID: 27401964]

[115] Miadoková E. Isoflavonoids - an overview of their biological activities and potential health benefits. Interdiscip Toxicol 2009; 2(4): 211-8.
[http://dx.doi.org/10.2478/v10102-009-0021-3] [PMID: 21217857]

[116] Qian Y, Guan T, Huang M, *et al.* Neuroprotection by the soy isoflavone, genistein, *via* inhibition of mitochondria-dependent apoptosis pathways and reactive oxygen induced-NF-κB activation in a cerebral ischemia mouse model. Neurochem Int 2012; 60(8): 759-67.
[http://dx.doi.org/10.1016/j.neuint.2012.03.011] [PMID: 22490611]

[117] Yang Z, Kulkarni K, Zhu W, Hu M. Bioavailability and pharmacokinetics of genistein: mechanistic studies on its ADME. Anticancer Agents Med Chem 2012; 12(10): 1264-80.
[http://dx.doi.org/10.2174/187152012803833107] [PMID: 22583407]

[118] Jacome LF, Gautreaux C, Inagaki T, *et al.* Estradiol and ERβ agonists enhance recognition memory, and DPN, an ERβ agonist, alters brain monoamines. Neurobiol Learn Mem 2010; 94(4): 488-98.
[http://dx.doi.org/10.1016/j.nlm.2010.08.016] [PMID: 20828630]

[119] Bean LA, Ianov L, Foster TC. Estrogen receptors, the hippocampus, and memory. Neuroscientist 2014; 20(5): 534-45.
[http://dx.doi.org/10.1177/1073858413519865] [PMID: 24510074]

[120] Bang OY, Hong HS, Kim DH, *et al.* Neuroprotective effect of genistein against beta amyloid-induced neurotoxicity. Neurobiol Dis 2004; 16(1): 21-8.
[http://dx.doi.org/10.1016/j.nbd.2003.12.017] [PMID: 15207258]

[121] Zeng Y, Li Y, Yang J, *et al.* Therapeutic Role of Functional Components in Alliums for Preventive Chronic Disease in Human Being. Evid Based Complement Alternat Med 2017; 2017: 9402849.
[http://dx.doi.org/10.1155/2017/9402849] [PMID: 28261311]

[122] Oi Y, Kawada T, Shishido C, *et al.* Allyl-containing sulfides in garlic increase uncoupling protein content in brown adipose tissue, and noradrenaline and adrenaline secretion in rats. J Nutr 1999; 129(2): 336-42.
[http://dx.doi.org/10.1093/jn/129.2.336] [PMID: 10024610]

[123] Sterea AM, Almasi S, El Hiani Y. The hidden potential of lysosomal ion channels: A new era of oncogenes. Cell Calcium 2018; 72: 91-103.
[http://dx.doi.org/10.1016/j.ceca.2018.02.006] [PMID: 29748137]

[124] Zhang L, Fang Y, Cheng X, *et al.* TRPML1 Participates in the Progression of Alzheimer's Disease by Regulating the PPARγ/AMPK/Mtor Signalling Pathway. Cell Physiol Biochem 2017; 43(6): 2446-56.
[http://dx.doi.org/10.1159/000484449] [PMID: 29131026]

[125] Li C, Li Q, Mei Q, Lu T. Pharmacological effects and pharmacokinetic properties of icariin, the major bioactive component in Herba Epimedii. Life Sci 2015; 126: 57-68.
[http://dx.doi.org/10.1016/j.lfs.2015.01.006] [PMID: 25634110]

[126] Song YH, Cai H, Zhao ZM, *et al.* Icariin attenuated oxidative stress induced-cardiac apoptosis by mitochondria protection and ERK activation. Biomed Pharmacother 2016; 83: 1089-94.
[http://dx.doi.org/10.1016/j.biopha.2016.08.016] [PMID: 27551754]

[127] Liu ZQ, Luo XY, Sun YX, *et al.* The antioxidative effect of icariin in human erythrocytes against free-

radical-induced haemolysis. J Pharm Pharmacol 2004; 56(12): 1557-62.
[http://dx.doi.org/10.1211/0022357044869] [PMID: 15563763]

[128] Liu B, Xu C, Wu X, *et al*. Icariin exerts an antidepressant effect in an unpredictable chronic mild stress model of depression in rats and is associated with the regulation of hippocampal neuroinflammation. Neuroscience 2015; 294: 193-205.
[http://dx.doi.org/10.1016/j.neuroscience.2015.02.053] [PMID: 25791226]

[129] Chen X, Mukwaya E, Wong MS, Zhang Y. A systematic review on biological activities of prenylated flavonoids. Pharm Biol 2014; 52(5): 655-60.
[http://dx.doi.org/10.3109/13880209.2013.853809] [PMID: 24256182]

[130] Guo J, Li F, Wu Q, Gong Q, Lu Y, Shi J. Protective effects of icariin on brain dysfunction induced by lipopolysaccharide in rats. Phytomedicine 2010; 17(12): 950-5.
[http://dx.doi.org/10.1016/j.phymed.2010.03.007] [PMID: 20382007]

[131] Li F, Gong QH, Wu Q, Lu YF, Shi JS. Icariin isolated from Epimedium brevicornum Maxim attenuates learning and memory deficits induced by d-galactose in rats. Pharmacol Biochem Behav 2010; 96(3): 301-5.
[http://dx.doi.org/10.1016/j.pbb.2010.05.021] [PMID: 20566405]

[132] Huang JH, Cai WJ, Zhang XM, Shen ZY. Icariin promotes self-renewal of neural stem cells: an involvement of extracellular regulated kinase signaling pathway. Chin J Integr Med 2014; 20(2): 107-15.
[http://dx.doi.org/10.1007/s11655-013-1583-7] [PMID: 24619236]

[133] Luo Y, Nie J, Gong QH, Lu YF, Wu Q, Shi JS. Protective effects of icariin against learning and memory deficits induced by aluminium in rats. Clin Exp Pharmacol Physiol 2007; 34(8): 792-5.
[http://dx.doi.org/10.1111/j.1440-1681.2007.04647.x] [PMID: 17600559]

[134] Angeloni C, Barbalace MC, Hrelia S. Icariin and Its Metabolites as Potential Protective Phytochemicals Against Alzheimer's Disease. Front Pharmacol 2019; 10: 271.
[http://dx.doi.org/10.3389/fphar.2019.00271] [PMID: 30941046]

[135] Gottlieb M, Leal-Campanario R, Campos-Esparza MR, *et al*. Neuroprotection by two polyphenols following excitotoxicity and experimental ischemia. Neurobiol Dis 2006; 23(2): 374-86.
[http://dx.doi.org/10.1016/j.nbd.2006.03.017] [PMID: 16806951]

[136] Ibarretxe G, Sánchez-Gómez MV, Campos-Esparza MR, Alberdi E, Matute C. Differential oxidative stress in oligodendrocytes and neurons after excitotoxic insults and protection by natural polyphenols. Glia 2006; 53(2): 201-11.
[http://dx.doi.org/10.1002/glia.20267] [PMID: 16206167]

[137] Lee MH, Cha HJ, Choi EO, *et al*. Antioxidant and cytoprotective effects of morin against hydrogen peroxide-induced oxidative stress are associated with the induction of Nrf-2□mediated HO-1 expression in V79-4 Chinese hamster lung fibroblasts. Int J Mol Med 2017; 39(3): 672-80.
[http://dx.doi.org/10.3892/ijmm.2017.2871] [PMID: 28204816]

[138] Zhang ZT, Cao XB, Xiong N, *et al*. Morin exerts neuroprotective actions in Parkinson disease models in *vitro* and in *vivo*. Acta Pharmacol Sin 2010; 31(8): 900-6.
[http://dx.doi.org/10.1038/aps.2010.77] [PMID: 20644549]

[139] Ji S, Li R, Wang Q, *et al*. Anti-H1N1 virus, cytotoxic and Nrf2 activation activities of chemical constituents from Scutellaria baicalensis. J Ethnopharmacol 2015; 176: 475-84.
[http://dx.doi.org/10.1016/j.jep.2015.11.018] [PMID: 26578185]

[140] Liang W, Huang X, Chen W. The Effects of Baicalin and Baicalein on Cerebral Ischemia: A Review. Aging Dis 2017; 8(6): 850-67.
[http://dx.doi.org/10.14336/AD.2017.0829] [PMID: 29344420]

[141] Li Y, Zhao J, Hölscher C. Therapeutic Potential of Baicalein in Alzheimer's Disease and Parkinson's Disease. CNS Drugs 2017; 31(8): 639-52.

[http://dx.doi.org/10.1007/s40263-017-0451-y] [PMID: 28634902]

[142] Oh SB, Park HR, Jang YJ, Choi SY, Son TG, Lee J. Baicalein attenuates impaired hippocampal neurogenesis and the neurocognitive deficits induced by γ-ray radiation. Br J Pharmacol 2013; 168(2): 421-31.
[http://dx.doi.org/10.1111/j.1476-5381.2012.02142.x] [PMID: 22891631]

[143] Liao KK, Wu MJ, Chen PY, *et al.* Curcuminoids promote neurite outgrowth in PC12 cells through MAPK/ERK- and PKC-dependent pathways. J Agric Food Chem 2012; 60(1): 433-43.
[http://dx.doi.org/10.1021/jf203290r] [PMID: 22145830]

[144] da Rocha Lindner G, Bonfanti Santos D, Colle D, *et al.* Improved neuroprotective effects of resveratrol-loaded polysorbate 80-coated poly(lactide) nanoparticles in MPTP-induced Parkinsonism. Nanomedicine (Lond) 2015; 10(7): 1127-38.
[http://dx.doi.org/10.2217/nnm.14.165] [PMID: 25929569]

[145] Haghi A, Azimi H, Rahimi R. A Comprehensive Review on Pharmacotherapeutics of Three Phytochemicals, Curcumin, Quercetin, and Allicin, in the Treatment of Gastric Cancer. J Gastrointest Cancer 2017; 48(4): 314-20.
[http://dx.doi.org/10.1007/s12029-017-9997-7] [PMID: 28828709]

[146] Bae HM, Kim SS, Cho CW, Yang DC, Ko SK, Kim KT. Antioxidant activities of ginseng seeds treated by autoclaving. J Ginseng Res 2012; 36(4): 411-7.
[http://dx.doi.org/10.5142/jgr.2012.36.4.411] [PMID: 23717144]

[147] Santín-Márquez R, Alarcón-Aguilar A, López-Diazguerrero NE, Chondrogianni N, Königsberg M. Sulforaphane - role in aging and neurodegeneration. Geroscience 2019; 41(5): 655-70.
[http://dx.doi.org/10.1007/s11357-019-00061-7] [PMID: 30941620]

[148] Park CE, Yun H, Lee EB, *et al.* The antioxidant effects of genistein are associated with AMP-activated protein kinase activation and PTEN induction in prostate cancer cells. J Med Food 2010; 13(4): 815-20.
[http://dx.doi.org/10.1089/jmf.2009.1359] [PMID: 20673057]

[149] Prasad K, Laxdal VA, Yu M, Raney BL. Antioxidant activity of allicin, an active principle in garlic. Mol Cell Biochem 1995; 148(2): 183-9.
[http://dx.doi.org/10.1007/BF00928155] [PMID: 8594422]

[150] Jeong JY, Cha HJ, Choi EO, *et al.* Activation of the Nrf2/HO-1 signaling pathway contributes to the protective effects of baicalein against oxidative stress-induced DNA damage and apoptosis in HEI193 Schwann cells. Int J Med Sci 2019; 16(1): 145-55.
[http://dx.doi.org/10.7150/ijms.27005] [PMID: 30662338]

CHAPTER 4

Therapeutic Potential of Vitamins in Parkinson's Disease

Prabhash Kumar Pandey[1,2,*], Shambhoo Sharan Tripathi[1], Jayant Dewangan[2], Ranjan Singh[3], Farrukh Jamal[4] and Srikanta Kumar Rath[2]

[1] *Department of Biochemistry, Faculty of Science, University of Allahabad, Prayagraj, Uttar Pradesh, India*

[2] *Genotoxicity lab, Division of Toxicology and Experimental Medicine, CSIR-Central Drug Research Institute, Lucknow, India*

[3] *Department of Biotechnology, Choithram College of Professional Studies, Indore, Madhya Pradesh, India*

[4] *Department of Biochemistry, Dr.Rammanohar Lohia Avadh University, Faizabad, Uttar Pradesh, India*

Abstract: Vitamins are naturally present in vegetables, spices, food supplements, and fruits. Vitamins can mitigate or prevent the pathophysiological phenomena involved in the progression of Parkinson's disease (PD). PD is a progressive and disabling syndrome that affects the person's quality of life by causing motor and non-motor disturbances and imposing an enormous burden on the caregivers. Oxidative stress (OS), neuroinflammation, mitochondrial dysfunction, and formation of free radicals are behind the PD. Various clinical scientific shreds of evidence explain the role of vitamins in the treatment of PD. Several cellular and animal-based experiments point out that proper intake of vitamins is helpful in PD treatment. The time, exact doses, and safety of regular consumption of these supplements still need to be explored more by the scientific community. A balanced diet with vitamins as supplements can boost up the current therapies used against the PD. Vitamins have the crucial antioxidant property that acts against the OS, thus helps in PD treatment. Through different molecular mechanisms, these vitamins protect dopaminergic neurons. There is a need for a cure against the PD. A promising approach to cure this disease by natural means, such as vitamins, has been focused throughout this chapter. In this book chapter, the authors collected the scientific evidence available throughout the various experimental platforms and literature related to the functional role of vitamins in the improvement of the clinical framework of PD patients.

Keywords: Antioxidant, Lewy bodies, Neuroprotection, Neurotoxicity, Oxidative stress, Parkinson's disease, Substantia nigra, Vitamins.

---

* **Corresponding author Prabhash Kumar Pandey:** Department of Biochemistry, Faculty of Science, University of Allahabad, Prayagraj, Uttar Pradesh, India; Tel: +91 6394252573; E-mail: prabhashpandey@allduniv.ac.in

Sachchida Nand Rai (Ed.)

INTRODUCTION

Among neurodegenerative diseases, Parkinson's disease (PD) in aged people is widespread. Both the environmental and genetic factors are involved in the onset and progression of PD [1]. Among these factors, improper function of mitochondria and oxidative stress (OS) are the main factors that trigger the PD [1, 2]. PD affects 1-2% of the population older than the age of sixty, along with Alzheimer's [3]. Retrogression of dopaminergic neurons in the substantia nigra pars compacta (SNpc) and the locus coeruleus mainly trigger PD [4]. Locus coeruleus causes psychological effects, and the SNpc controls motor activity. The presence of Lewy neuritis and Lewy bodies (LB) in the brain represents the progression of PD [5]. Protein agglomerations such as parkin, alpha-synuclein, and some other proteins form the LB. In the different parts of the brain, these LB are present in the cytoplasm of the dead neurons [6]. Agglomerations of proteins form the large insoluble fibrils, which play a significant role in PD [7].

Different molecular mechanisms are involved in PD's pathophysiology, and they are all linked to each other [6]. In PD patients, many non-motor symptoms like olfactory dysfunction, sleep disorders, cognitive decline, constipation, and autonomic symptoms appear before the motor symptoms [8, 9]. OS is the leading cause behind PD's pathophysiology; therefore, those strategies regulating redox balance must come into practice for the treatment of PD [5].

Neuronal mitochondrial dysfunction is behind the pathophysiology of PD. Abnormal production of α-synuclein can cause mitochondrial dysfunction, leading to OS and neuronal degeneration in PD patients [10, 11].

To hold back the motor disorders, restoration of the dopamine level in the striatum is the turning point in PD treatment [12]. Treatment with the levodopa helps in peripheral dopamine metabolism. Levodopa smoothens the bioavailability of dopamine, thus lowering the motor complications [13, 14].

Vitamins have beneficial anti-oxidative as well as gene expression regulating tributes [15]. These marvelous properties of vitamins show their usefulness against the PD. Vitamins counteract or mitigate PD's pathophysiological phenomena, thus improving PD patients' cognitive functions and learning process [16]. Several clinical studies suggest that vitamins slow down PD progression in human beings; therefore, these nutrients can pretend as an adjuvant in the treatment of PD [17]. PD treatment is a challenging task, and current strategies to treat the PD can only relax the clinical symptoms, and they are not capable of stopping the PD progression. In this book chapter, the biological interconnections between PD and vitamins and their therapeutic role in PD treatment have been discussed in detail.

ROLE OF OXIDATIVE STRESS IN THE PATHOGENESIS OF PD

Production of reactive oxygen and nitrogen like oxidative entities damage the imbalance between the antioxidant and oxidative systems, and as a result, OS happens [18]. OS actively participates in many physiological events of the organism. OS facilitates both the process of xenobiotic metabolism and the production of biologically active substances. It also kills the pathogenic microorganisms *via* phagocytosis [15]. OS can modulate the nucleic acid, takes part in the protein denaturation, damages the cellular membrane, *etc* [15]. Various studies suggest that reactive oxygen species (ROS) generated by the OS causes the death of neurons, and it is the main culprit behind PD's onset and progression [19, 20]. The respiratory chain of mitochondria is the leading site of ROS [21]. A clinical study by Schapira *et al.* [14, 22] shows that in the substantia nigra of PD patients, mitochondrial dysfunction happens. Accumulation of iron in substantia nigra of PD patients can trigger molecular oxygen and hydrogen peroxide production through the Haber-Weiss reaction [23]. Hydrogen peroxide produces hydroxyl radical, which is highly toxic. Severe oxidative damages occur due to this radical in cellular components [23]. Oxidation of dopamine and its metabolites reduces the activity of mitochondrial complex I [24]. Reduced glutathione boosts the generation of ROS [25]. Dopaminergic neurons become more susceptible after the production of their metabolites.

Accumulation of proteins, nonproper functioning of mitochondria, and DNA damage are severe intracellular events. OS initiates these processes, which lead to neuronal loss and PD [15]. Targeting these events can be beneficial in the treatment of PD. Vitamins could be a better way to treat the PD because these vitamins are the key to several biochemical pathways. In almost all tissues, vitamins act as enzyme cofactors [26]. These vitamins improve the nervous and immune system, regulate metabolism, and control cell growth and cell division events [26].

THEATRICAL ROLE OF NEUROINFLAMMATION IN THE PATHOGENESIS OF PD

In the pathogenesis of PD along with the OS, neuroinflammation also participates. It enhances the microglial activity and increases the production of pro-inflammatory and toxic mediators [27]. Insoluble fibrils trigger the activation of immunoinflammatory cells with the help of α-synuclein. These activated immune cells release the neurotoxic mediators and several messenger cells, such as chemokines and cytokines. These mediators initiate the degeneration process in dopaminergic neurons [28]. Insoluble fibrils are toxic, and they directly affect the operation of neurodegeneration and neuronal cell death. They execute these processes by activating the pro-inflammatory processes [29]. The migration of peripheral inflammatory and immune cells happens through the faulty blood-brain barrier. These crossed cells cause neuroinflammation directly or with the help of neuronal and glial cells [27]. Taking anti-inflammatory substances can decrease or suppress neuroinflammation, thus improving PD symptoms and slowing down the expansion of this disease in PD patients.

NEUROPROTECTIVE ROLES OF VITAMINS IN THE TREATMENT OF PARKINSONISM

Neuroprotective Role of Vitamin B in PD

Vitamin B presents abundantly in pulses, potato, whole grains, banana, and in all types of meats. Vitamin B3 is also known as niacin. Its active form nicotinamide is present in fish, wheat, and meat, while in vegetables, it is present in a lesser amount [30]. The deficiency of vitamin B3 causes several issues like dermatitis, depression, diarrhea, and pellagra [31]. Nicotinamide acts as a precursor of coenzymes NADH and NADPH [30]. These coenzymes participate in the production of energy currency adenosine triphosphate [30]. In recent times the use of vitamin B3 is getting more attention in the treatment of PD.

A study depicts that vitamin B3 at low doses acts as a neuroprotective and antioxidant agent. Vitamin B3 causes neurotoxicity at high doses in the dopaminergic neurons [32]. High doses of niacin increase the production of 1-methyl nicotinamide (MNA) that is harmful to PD patients [33]. More than 20mM doses of vitamin B3 triggers cytotoxic events and leads cells towards death [34]. To determine the exact beneficial dose of vitamins that can help in the PD treatment still want some more research. Griffin *et al.* [34] reported that *in vitro* 10mM dose of nicotinamide differentiates the embryonic stem cells into neurons.

Both mitochondrial dysfunction and failure of cellular energy represent the pathophysiology of PD. Nicotinamide plays a vital role in the biosynthesis of nicotinamide adenine dinucleotide (NAD) [35]. An essential cofactor, NADH, helps in the functioning of tyrosine hydroxylase enzymes [36]. This hydroxylase enzyme produces dopamine by hydroxylating the tyrosine. In PD patients, the NADH deficiency is prevalent in the mitochondrial complex I, and NADH is crucial for ATP synthesis [36]. Both in the animal models and PD patients, the above function does not happen smoothly [14, 37, 38]. Rotenone treated PC12 cell line confirms that after the association of Nicotinamide with mononucleotide [(NMN) at a concentration of 0.1nM or 1mM)] enhances the survival rate of PC12 cells [39].

NMN generates important precursors of NAD+, and it increases the intracellular level of ATP and NAD+ in the cellular model of PD [39]. Nicotinamide inhibits the process of OS; thus, it is an ideal vitamin for PD treatment. A cotreatment study with nicotinamide reveals that administration of 500mg/kg nicotinamide in mice before the subacute doses of 1-methyl-4-phenyl-1,2,3,6-tetrahydropyridine-[(MPTP-) 30 mg/kg/d for five days] improves the locomotor activity significantly in comparison to the individual agent administration with the MPTP- [40]. Administration of nicotinamide withdraws the MPTP induced depletion of dopamine dramatically [41]. Nicotinamide pre-treatment inhibits MPTP- induced activities of NOS and LDH and finally reduces the initiation of OS [41]. Sirtuins take part in many vital biological phenomena, and NAD+ dependent enzyme deacetylases produce theme [40].

Some studies like Liu *et al.* [41] suggest the considerable role of sirtuins in the MPTP-induced PD mouse-models. Liu *et al.* [41] indicated that after the deletion of sirtuins, degeneration of nigrostriatal dopaminergic neurons occurs in the substantia nigra. Not only in PD, vitamins also play a protective role in Alzheimer's disease [42 - 44]. This event ruins the anti-oxidative properties of mitochondria in the mouse model of PD [40]. The Above studies create a promising hope for the treatment of PD. With this, some controversial studies regarding the neuroprotective role of niacin point out towards more need for clinical studies for the exact clarification regarding the role of niacin in the PD treatment [45]. Many clinical studies related to the neuroprotective role of vitamin B3 show that a diet containing a high level of niacin lowers the risk of PD [46, 47]. In idiopathic PD patients, an oral dose (500mg) of niacin used consecutively for three months reduces the bradykinesia and rigidity. Although the treatment was related to hypertriglyceridemia and when the oral dose discontinues by the patient due to the skin rash and nightmares issues, the stiffness and bradykinesia revived [48].

In the PD treatment role of vitamin B is directly depends on the part of homocysteine. The methionine cycle generates neurotoxic substance homocysteine that causes neurodegenerative diseases. It has the neurotoxic property that leads to the PD and other neurodegenerative disorders [49, 50]. In PD patients, a high level of homocysteine is present compared to the healthy persons of the same age groups [51, 52]. Homocysteine promotes the cell death process of dopaminergic neurons in PD patients [53 - 55]. Vitamin B act as a cofactor in the synthesis of essential amino-acid methionine from the homocysteine. Therefore, a high and therapeutic intake of vitamin B protects against the PD by reducing the homocysteine levels in the plasma. Vitamin B6 has antioxidant effects, reducing the PD risk by participating in the dopamine synthesis [56].

Usability of Vitamin C in PD Treatment

Vitamin C is present in various tissues in plenty. This essential nutrient supplement is found in the livers of the animals, fresh fruits (especially in citrus fruits and paprika), and vegetables abundantly [6]. A reduced form of vitamin C, known as ascorbate, supports tissue regeneration, and maintains homeostasis in healthy tissues and organs. It scavenges ROS emerges by the process of neuronal metabolism and synaptic activity. Ascorbate facilitates the neuromodulatory functions in the brain. A relationship exists between the low level of ascorbate in serum and neurodegenerative disorders. In neurodegenerative diseases, ascorbic acid provides the anti-oxidative defense and modifies the astrocytic and neuronal metabolism and [57].

To fight against neuromodulation and OS, the presence of intracellular ascorbate is crucial. In PD patients, Its deficiency in brain areas can exacerbate metabolic failure, tissue damage, and redox imbalance. Vitamin C (ascorbic acid) is soluble in water, and it has two molecular subforms oxidized [dehydroascorbic acid (DHA)] and reduced [ascorbic acid (AA)], respectively [15]. Ascorbic acid has antioxidant properties, and this attribute is useful in the scavenging process of free radicals, in the reduction of the OS, and lipid peroxidation [58]. Ascorbic acid is also beneficial in the regeneration of the other antioxidants [57].

The downstream metabolism process of dopamine generates ROS and RNS. These products trigger the gathering of abnormal proteins in PD [59]. Some fundamental properties of vitamin C could be used in the PD treatment, as the presence of it being noticed in those areas rich in neurons [60, 61]. Vitamin C transporter type 2 (SVCT2) carries vitamin C to the brain, and two transporters, glucose transporter type 1 (GLUT1) and glucose transporter type 3 (GLUT3), take the DHA in the brain [62, 63].

Seitz *et al.* [64] report that vitamin C increases the dihydroxyphenylalanine (DOPA) production when it is incubated with a cell line SK-N-SH (human neuroblastoma cell line) in a dose-dependent manner. A threefold enhancement of tyrosine hydrolase expression was noticed after the incubation with the ascorbic acid [64]. Two research outputs indicate that Vitamin C has a protective capacity against the levodopa and MPTP mediatedneurotoxicity [65, 66]. In aged persons,

Vitamin C controls the absorption of levodopa by reducing its bioavailability [67]. An effective strategy could be applied by using the available drugs of PD in combination with vitamin C [68]. He *et al.* [69] report that ascorbic acid application increased an *in vitro* differentiation of embryonic midbrain neural stem cells into midbrain dopaminergic neurons. A 10 fold dopaminergic differentiation was noticed in the central nervous system's precursor cells when treated with vitamin C [70]. A study also reports that vitamin C triggers the differentiation of precursor cells of CNS into the neuronal and glial cells [71].

Ten-eleven translocation one methylcytosine dioxygenase 1 (TET1) catalyzed dopaminergic phenotypic gene promoters produces 5 hydroxymethylcytosine (5HMC) [[72, 69]. Vitamin C increases the production of 5HMC [72]. On the other hand, histone H3K27 demethylase (JMJD3) also catalyzes the dopaminergic phenotype gene promoters, and when vitamin C is incorporated, it induces the loss of H3K27m3. Experimentally it was shown that the epigenetic role of vitamin C is connected with the development of dopaminergic neurons in the midbrain [72].

A clinical study involving 1036 patients of PD suggests that the intake of vitamin C is beneficial in PD treatment because it lowers the risk of PD. Still, for a four-year lag analysis, its effectiveness is not valid [73]. There are several other clinical studies performed by scientists who indicate controversially that ascorbic acid intake does not control the PD [74, 75]. It might be assumed that this controversy is connected with the patients' timing of vitamin C uptake.

Role of Vitamin E in the Treatment of Pd

Vitamin E regulates the physiological performances, gene expression and takes part in the immune and cognitive functions [76 - 78]. Anemia, ataxia, and peripheral neuropathy are the main clinical characteristics of vitamin E deficiency in the organism [79, 80]. Lack of vitamin E is rare, and its defect is reported in premature babies and infants [15].

There are two subgroups named tocotrienols and tocopherols present in this vitamin. These subgroups divide into four lipophilic molecules (α-, β-, γ-, and δ-tocotrienol and α-, β-, γ-, and δ-tocopherol, respectively). Both the subgroups can be differentiated primarily by the side chains [15].

Tocotrienols have an unsaturated isoprenoid side chain; however, tocopherol contains a saturated side chain, *i.e.,* phytyl tail. Tocotrienol has more anti-oxidative properties than tocopherol due to the presence of an unsaturated side chain. This unsaturated side chain helps in electron acceptance and molecular mobility through the lipid membrane [15].

Cadet *et al.* [81] reported that in rats, pre-treatment with tocopherol weakens the biochemical irregularities and circling behaviors induced by the striatum injections of 6-hydroxydopamine [81]. Two other studies also report the neuroprotective role of vitamin E against rotenone and 6-hydroxydopamine induced neurotoxicity [82, 83]. Vitamin E cannot wholly protect dopaminergic neurons after intoxication with MPTP- in the mouse model of PD [84, 85]. Vitamin E inhibits apoptosis and OS; thus, it works as a neuroprotective agent.

Tocotrienol actively participates in signal transduction of estrogen beta receptors and OS [86].

Nakaso *et al.* [87] report that γ-tocotrienol/δ-tocotrienol exerts a neuroprotective role in the SY5Y cells by inhibiting the neurotoxicity triggered by the MPP+. ERβ-PI3K/Akt signaling pathways are behind the neuroprotective role of γ-tocotrienol/δ-tocotrienol. δ-tocotrienol administration reduces dopaminergic neurons in the compact area of the substantia nigra [87]. ER inhibitors weaken the neuroprotective effects of δ-tocotrienol. These findings strengthen the therapeutic potential of vitamin E in the treatment of PD.

A clinical study based on the broader community based reveals that 10 mg/day intake of dietary vitamin E is beneficial in the control individuals of PD [88]. Long term vitamin E use minimizes the administration of levodopa in PD patients [89]. A high dose (up to 2000 IU/day) administration of vitamin E enhances its level in the cerebrospinal fluid [90]. Other studies suggest that vitamin E present uniformly in the serum, brains, and CSF of PD and control individuals [91 - 93]. A Deprenyl and Tocopherol Antioxidative Therapy of Parkinsonism (DATATOP) based clinical trial is performed to explore the long term performance of both vitamin E and deprenyl in PD treatment. The study reveals that the performance of vitamin E is inferior, while deprenyl delayed the levodopa application. Deprenyl halts the growth of functional disorders and maintain motor symptoms. Two other group studies also did not found the therapeutic correlations between the PD and vitamin E [17, 94].

The potent antioxidant properties of vitamin E are the only factor that shows its significance in the PD treatment. The exact protective mechanism of vitamin E against the PD is not very clear to date; therefore, more studies come into practice to determine the effectiveness of vitamin E in PD treatment.

Role of Vitamin D as a Neuroprotectant in the Pd Treatment

Vitamin D exists in two forms vitamin D2 and D3 [15]. The D3 form of this vitamin is produced endogenously by the skin cells when exposed to Sun's UV-B rays [15]. Vitamin D3 is naturally available in cod liver oil, salmon, tuna, fat cheese, carp, and mushrooms. It acts as a neurosteroid and is crucial for the brain's function and development [95]. Several studies reveal that it heightens the antioxidant concentrations and act as a neurotrophic factor. Vitamin D protects against excitotoxicity [96].

Vitamin D plays a substantive role in maintaining skeletal health and calcium homeostasis [15]. Both the forms of vitamin D are present in an inactive form, and after the hydroxylation process, they converted into a more active form 1,25-dihydroxy vitamin D3 [97, 98]. This dynamic form secretes in the blood through the kidney and further binds to the nuclear vitamin D receptor and regulates the gene expression [99, 100].

Vitamin D affects many regulatory processes like anti-oxidative stress, immunomodulation, cell proliferation, and differentiation. These processes are physiologically as well as pathologically relevant [101 - 103]. Vitamin D3 halts free radical generation and inhibits the process of OS [104]. It reduces the neurotoxicity by promoting the signaling pathways of autophagy and slows the endothelial dysfunction seen in PD patients [105]. Its deficiency affects all age groups of human beings; children may have rickets and old-age people with osteomalacia [15]. Some other disease comes into existence in the lack of vitamin D such as diabetes mellitus, cardiovascular diseases, cancers, muscle weakness, and multiple sclerosis [106].

The human vitamin D receptor gene (VDR) has three alternative 5' noncoding exons and eight coding exons. All belong to the intranuclear receptors superfamily. VDR occupies on chromosome number 12 and spans on it over 105 kb [107]. ApaI, TaqI, FokI, and BsmI are the most studied biallelic polymorphic sites of VDR. Many studies report that there are some relations between these biallelic sites and PD. The bb genotype of VDR is presented at high levels in PD patients as compared to the control individuals [108]. The bb genotype of VDR is less common in PD patients having tremors than in those patients who have gait difficulty and postural instability [108]. The FokI genotype of VDR is a concern with the reduction in cognitive properties and severity of PD [109, 110]. With this, a study showed that FokI and BsmI sites of VDR are associated with the risk of PD [111].

In VDR-knockout mice, motor and muscle impairment was reported, which broadly changed motor behavior; it points out the importance of vitamin D in the pathogenesis of PD [108]. Glial cell line-derived neurotrophic factor (GDNF) plays a crucial role in the existence and maintenance of dopaminergic neurons. It inhibits the activation of microglial cells [112]. The endogenous expression of GDNF protects the dopaminergic neurons from the inflammation caused by the immune system [113]. Both *in vitro* and *in vivo* studies show that 1,25-(OH)2-D3 can accelerate the endogenous expression of GDNF, which inhibits the activation of the glial cell for neuroprotection [113 - 115]. Vitamin D3 protects dopaminergic neurons against the 6-hydroxydopamine induced neurotoxicity. In PD rat models, vitamin D3 alleviates motor performance [28]. This study reveals that Vitamin D scavenges free radicals, inhibits OS events, and reduces ROS production [116].

The nonproper function of endothelial cells in PD patients is related to the lower presence of vitamin D in them [105]. Several clinical and epidemiological studies reveal that individuals with high vitamin D content in their serum are at a lower risk of PD. These findings strengthen the fact about the beneficial role of vitamin D in the treatment of PD [117, 118].

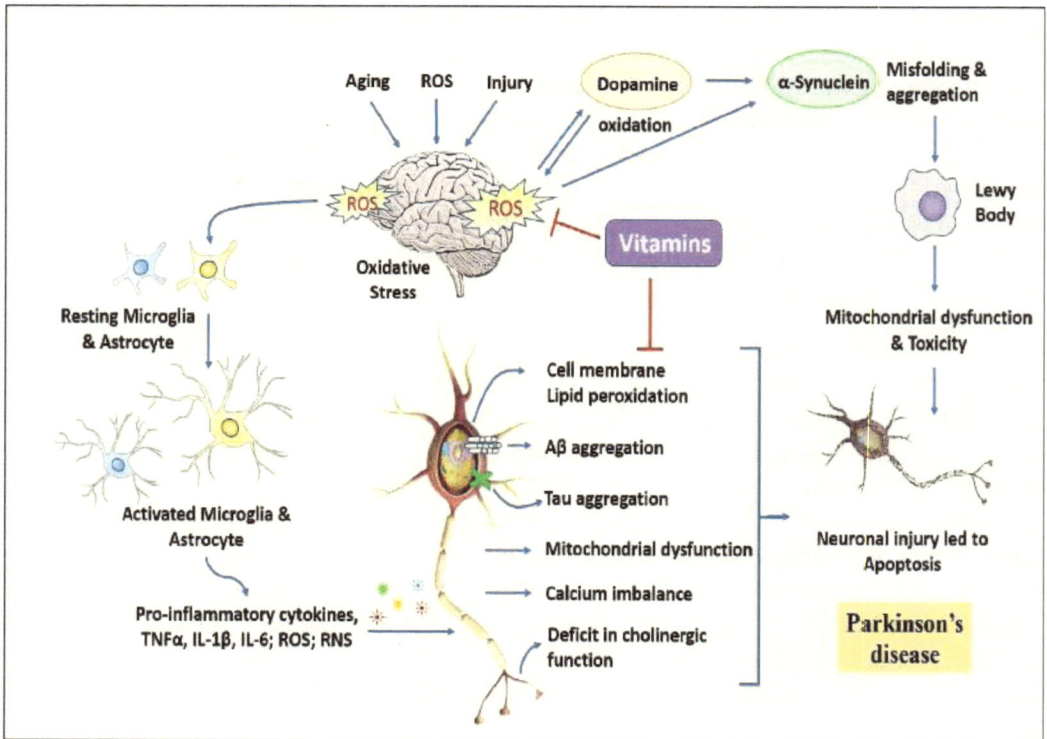

Fig. (1). An overview of different molecular mechanisms involved in the pathophysiology of Parkinson's disease and the roles of vitamins in the PD treatment.

A study related to outdoor workers suggests that these persons are at a lower risk of PD. The finding behind this study is that the D3 form of vitamin D is produced in these outdoor workers because they are exposed more to the sunlight [119]. This study depends on a large number of Danish people having both the control and PD patients. An ecologic survey of the France people also suggested the beneficial role of vitamin D in the treatment of PD [120]. Sunlight exposed individuals with a lower level of 25-hydroxy vitamin D are at the risk of PD, but when an adequate amount of 25-hydroxy vitamin D is present in the serum, it controls the progression of PD [121]. More studies can determine the exact relationship between PD and vitamin D. Fig. (**1**) exhibits different molecular mechanisms involved in the pathophysiology of Parkinson's disease along with the role of vitamins in the PD treatment.

CONCLUSION AND FUTURE DIRECTIONS

After summarizing the beneficial role of different vitamins in the treatment of PD, we find that the intake of vitamins can lower the risk of PD in the organism. Some attributes of vitamins such as bioavailability, chemical properties, absorption rate, and time of intake may affect their role in the PD treatment. Fat-soluble vitamins E and D have raised hope in the treatment of PD. Many reports are also present in the scientific literature that vitamin E is ineffective in PD treatment. Vitamin D lowers the risk of bone fracture in PD patients and improves their conditions toward the fight against the PD. Timely intake of vitamins plays a crucial role in PD treatment. Vitamins are natural molecules, and these compounds do not express their benefits in a short period against the slow-running PD. Therefore, intake of vitamins by PD patients over a long time duration exert several beneficial outcomes.

In this book chapter, we also focus on the role and properties of water-soluble vitamins (vitamin B3 and vitamin C) in PD treatment. We conclude that more clinical studies should come into practice for the exact evaluation of both the harmful and beneficial role of these vitamins in PD treatment. Vitamin C helps in the absorption of levodopa in aged people suffering from PD. More clinical studies can reveal the exact serum level of vitamin C and its safe supplementation in PD patients. Several studies were reported by the scientific groups who depicted the role of these vitamins in the treatment of PD. However, there is still a lack of more clinical trials by which we would confidentially say that vitamin intake reduces the onset and progression of PD.

CONSENT FOR PUBLICATION

Not applicable.

CONFLICT OF INTEREST

The authors declare no conflict of interest, financial or otherwise.

ACKNOWLEDGEMENTS

The corresponding author expresses his gratitude to the University Grants Commission, New Delhi (UGC-DSKPDF No.F.4-2/2006 BSR/BL/17-18/0442) and, Department of Science and Technology SERB (PDF/2015/000033)for providing financial assistance. SSTacknowledgesUniversity Grants Commission, New Delhi (UGC-DSKPDF No.F.4-2/2006 BSR/BL/17-18/381), and JD are also thankful to University Grants Commission, New Delhi for providing Junior and Senior research fellowships.

REFERENCES

[1]   Hattingen E, Magerkurth J, Pilatus U, *et al.* Phosphorus and proton magnetic resonance spectroscopy demonstrates mitochondrial dysfunction in early and advanced Parkinson's disease. Brain 2009; 132(Pt 12): 3285-97.
[http://dx.doi.org/10.1093/brain/awp293] [PMID: 19952056]

[2]   Parker WD Jr, Parks JK, Swerdlow RH. Complex I deficiency in Parkinson's disease frontal cortex. Brain Res 2008; 1189: 215-8.
[http://dx.doi.org/10.1016/j.brainres.2007.10.061] [PMID: 18061150]

[3]   von Campenhausen S, Bornschein B, Wick R, *et al.* Prevalence and incidence of Parkinson's disease in Europe. Eur Neuropsychopharmacol 2005; 15(4): 473-90.
[http://dx.doi.org/10.1016/j.euroneuro.2005.04.007] [PMID: 15963700]

[4]   Braak H, Rüb U, Gai WP, Del Tredici K. Idiopathic Parkinson's disease: possible routes by which vulnerable neuronal types may be subject to neuroinvasion by an unknown pathogen. J Neural Transm (Vienna) 2003; 110(5): 517-36.
[http://dx.doi.org/10.1007/s00702-002-0808-2] [PMID: 12721813]

[5]   Schirinzi T, Martella G, Imbriani P, *et al.* Dietary Vitamin E as a Protective Factor for Parkinson's Disease: Clinical and Experimental Evidence. Front Neurol 2019; 10: 148.
[http://dx.doi.org/10.3389/fneur.2019.00148] [PMID: 30863359]

[6]   Ciulla M, Marinelli L, Cacciatore I, Stefano AD. Role of Dietary Supplements in the Management of Parkinson's Disease. Biomolecules 2019; 9(7): 271.
[http://dx.doi.org/10.3390/biom9070271] [PMID: 31295842]

[7]   Xilouri M, Brekk OR, Stefanis L. α-Synuclein and protein degradation systems: a reciprocal relationship. Mol Neurobiol 2013; 47(2): 537-51.
[http://dx.doi.org/10.1007/s12035-012-8341-2] [PMID: 22941029]

[8]   Khoo TK, Yarnall AJ, Duncan GW, *et al.* The spectrum of nonmotor symptoms in early Parkinson disease. Neurology 2013; 80(3): 276-81.
[http://dx.doi.org/10.1212/WNL.0b013e31827deb74] [PMID: 23319473]

[9]   Postuma RB, Aarsland D, Barone P, *et al.* Identifying prodromal Parkinson's disease: pre-motor disorders in Parkinson's disease. Mov Disord 2012; 27(5): 617-26.

[http://dx.doi.org/10.1002/mds.24996] [PMID: 22508280]

[10]   Kaushik S, Cuervo AM. Proteostasis and aging. Nature Medicine 2015; Vol. 21: pp. 1406-15.

[11]   Wales P, Pinho R, Lázaro DF, Outeiro TF. Limelight on alpha-synuclein: pathological and mechanistic implications in neurodegeneration. J Parkinsons Dis 2013; 3(4): 415-59. [http://dx.doi.org/10.3233/JPD-130216] [PMID: 24270242]

[12]   Lewitt PA, Fahn S. Levodopa therapy for Parkinson disease: A look backward and forward. Neurology 2016; Vol. 86: S3-S12.

[13]   Fox SH, Katzenschlager R, Lim SY, Barton B, *et al.* International Parkinson and movement disorder society evidence-based medicine review: Update ontreatments for the motor symptoms of Parkinson's disease. Movement Disorders 2018; 33: 1248-66.

[14]   Schapira AH, Cooper JM, Dexter D, Jenner P, Clark JB, Marsden CD. Mitochondrial complex I deficiency in Parkinson's disease. Lancet 1989; 1(8649): 1269. [http://dx.doi.org/10.1016/S0140-6736(89)92366-0] [PMID: 2566813]

[15]   Zhao X, Zhang M, Li C, Jiang X, Su Y, Zhang Y. Benefits of Vitamins in the Treatment of Parkinson's Disease. Oxidative Medicine and Cellular Longevity 2019. vol. 2019, Article ID 9426867, 14 pages.

[16]   Olasehinde T, Oyeleye SI, Ogunsuyi OB, *et al.* Functional Foods in the management of Neurodegenerative Diseases.Functional Foods: Unlocking the Medicine in Foods. Memphis, TN, USA: GracelandPrints 2017; pp. 72-81.

[17]   Zhang SM, Hernán MA, Chen H, Spiegelman D, Willett WC, Ascherio A. Intakes of vitamins E and C, carotenoids, vitamin supplements, and PD risk. Neurology 2002; 59(8): 1161-9. [http://dx.doi.org/10.1212/01.WNL.0000028688.75881.12] [PMID: 12391343]

[18]   Sies H, Berndt C, Jones DP. Oxidative Stress. Annu Rev Biochem 2017; 86(1): 715-48. [http://dx.doi.org/10.1146/annurev-biochem-061516-045037] [PMID: 28441057]

[19]   Kalia LV, Lang AE. Parkinson's disease. Lancet 2015; 386(9996): 896-912. [http://dx.doi.org/10.1016/S0140-6736(14)61393-3] [PMID: 25904081]

[20]   Blesa J, Trigo-Damas I, Quiroga-Varela A, Jackson-Lewis VR. Oxidative stress and Parkinson's disease. Front Neuroanat 2015; 9: 91. [http://dx.doi.org/10.3389/fnana.2015.00091] [PMID: 26217195]

[21]   Murphy MP. How mitochondria produce reactive oxygen species. Biochem J 2009; 417(1): 1-13. [http://dx.doi.org/10.1042/BJ20081386] [PMID: 19061483]

[22]   Schapira AHV. Monoamine Oxidase B Inhibitors for the Treatment of Parkinson's Disease. CNS Drugs 2011; Vol. 25: pp. 1061-71.

[23]   Hare DJ, Double KL. Iron and dopamine: a toxic couple. Brain 2016; 139(Pt 4): 1026-35. [http://dx.doi.org/10.1093/brain/aww022] [PMID: 26962053]

[24]   Gluck MR, Zeevalk GD. Inhibition of brain mitochondrial respiration by dopamine and its metabolites: implications for Parkinson's disease and catecholamine-associated diseases. J Neurochem 2004; 91(4): 788-95. [http://dx.doi.org/10.1111/j.1471-4159.2004.02747.x] [PMID: 15525332]

[25]   Sian J, Dexter DT, Lees AJ, *et al.* Alterations in glutathione levels in Parkinson's disease and other neurodegenerative disorders affecting basal ganglia. Ann Neurol 1994; 36(3): 348-55. [http://dx.doi.org/10.1002/ana.410360305] [PMID: 8080242]

[26]   Mikkelsen K, Stojanovska L, Tangalakis K, Bosevski M, Apostolopoulos V. Cognitive decline: A vitamin B perspective. Maturitas 2016; 93: 108-13. [http://dx.doi.org/10.1016/j.maturitas.2016.08.001] [PMID: 27574726]

[27]   Kempuraj D, Thangavel R, Natteru P, Selvakumar G, *et al.* Neuroinflammation Induces

Neurodegeneration HHS Public Access. J Neurol Neurosurg Spine 2016; 1: 1-15.

[28] Wang JY, Wu JN, Cherng TL, *et al.* Vitamin D(3) attenuates 6-hydroxydopamine-induced neurotoxicity in rats. Brain Res 2001; 904(1): 67-75.
[http://dx.doi.org/10.1016/S0006-8993(01)02450-7] [PMID: 11516412]

[29] Grimmig B, Morganti J, Nash K, Bickford PC. Immunomodulators as therapeutic agents in mitigating theprogression of Parkinson's disease. Brain Sci 2016; 6(4): 41.
[http://dx.doi.org/10.3390/brainsci6040041] [PMID: 27669315]

[30] Gehring W. Nicotinic acid/niacinamide and the skin. J Cosmet Dermatol 2004; 3(2): 88-93.
[http://dx.doi.org/10.1111/j.1473-2130.2004.00115.x] [PMID: 17147561]

[31] Surjana D, Damian DL. Nicotinamide in dermatology and photoprotection. Skinmed 2011; 9(6): 360-5.
[PMID: 22256624]

[32] Williams A, Ramsden D. Nicotinamide: a double edged sword. Parkinsonism Relat Disord 2005; 11(7): 413-20.
[http://dx.doi.org/10.1016/j.parkreldis.2005.05.011] [PMID: 16183323]

[33] Aoyama K, Matsubara K, Kondo M, *et al.* Nicotinamide-N-methyltransferase is higher in the lumbar cerebrospinal fluid of patients with Parkinson's disease. Neurosci Lett 2001; 298(1): 78-80.
[http://dx.doi.org/10.1016/S0304-3940(00)01723-7] [PMID: 11154840]

[34] Griffin SM, Pickard MR, Orme RP, Hawkins CP, Fricker RA. Nicotinamide promotes neuronal differentiation of mouse embryonic stem cells *in vitro*. Neuroreport 2013; 24(18): 1041-6.
[http://dx.doi.org/10.1097/WNR.0000000000000071] [PMID: 24257250]

[35] Imai S. Nicotinamide phosphoribosyltransferase (Nampt): a link between NAD biology, metabolism, and diseases. Curr Pharm Des 2009; 15(1): 20-8.
[http://dx.doi.org/10.2174/138161209787185814] [PMID: 19149599]

[36] Pearl SM, Antion MD, Stanwood GD, Jaumotte JD, Kapatos G, Zigmond MJ. Effects of NADH on dopamine release in rat striatum. Synapse 2000; 36(2): 95-101.
[http://dx.doi.org/10.1002/(SICI)1098-2396(200005)36:2<95::AID-SYN2>3.0.CO;2-U]   [PMID: 10767056]

[37] Mizuno Y, Ohta S, Tanaka M, *et al.* Deficiencies in complex I subunits of the respiratory chain in Parkinson's disease. Biochem Biophys Res Commun 1989; 163(3): 1450-5.
[http://dx.doi.org/10.1016/0006-291X(89)91141-8] [PMID: 2551290]

[38] Nicklas WJ, Vyas I, Heikkila RE. Inhibition of NADH-linked oxidation in brain mitochondria by 1-methyl-4-phenyl-pyridine,   a   metabolite   of   the   neurotoxin,   1-methyl-4-phenyl-1,2-5,6-tetrahydropyridine. Life Sci 1985; 36(26): 2503-8.
[http://dx.doi.org/10.1016/0024-3205(85)90146-8] [PMID: 2861548]

[39] Lu L, Tang L, Wei W, *et al.* Nicotinamide mononucleotide improves energy activity and survival rate in an *in vitro* model of Parkinson's disease. Exp Ther Med 2014; 8(3): 943-50.
[http://dx.doi.org/10.3892/etm.2014.1842] [PMID: 25120628]

[40] Yang L, Ma X, He Y, *et al.* Sirtuin 5: a review of structure, known inhibitors and clues for developing new inhibitors. Sci China Life Sci 2017; 60(3): 249-56.
[http://dx.doi.org/10.1007/s11427-016-0060-7] [PMID: 27858336]

[41] Liu L, Peritore C, Ginsberg J, Shih J, Arun S, Donmez G. Protective role of SIRT5 against motor deficit and dopaminergic degeneration in MPTP-induced mice model of Parkinson's disease. Behav Brain Res 2015; 281: 215-21.
[http://dx.doi.org/10.1016/j.bbr.2014.12.035] [PMID: 25541039]

[42] Bhatti AB, Usman M, Ali F, Satti SA. Vitamin Supplementation as an Adjuvant Treatment for Alzheimer's Disease. J Clin Diagn Res 2016; 10(8): OE07-11.
[http://dx.doi.org/10.7860/JCDR/2016/20273.8261] [PMID: 27656493]

[43]   Lloret A, Esteve D, Monllor P, Cervera-Ferri A, Lloret A. The Effectiveness of Vitamin E Treatment in Alzheimer's Disease. Int J Mol Sci 2019; 20(4): 879.
[http://dx.doi.org/10.3390/ijms20040879] [PMID: 30781638]

[44]   Ford AH, Almeida OP. Effect of Vitamin B Supplementation on Cognitive Function in the Elderly: A Systematic Review and Meta-Analysis. Drugs Aging 2019; 36(5): 419-34.
[http://dx.doi.org/10.1007/s40266-019-00649-w] [PMID: 30949983]

[45]   Johnson CC, Gorell JM, Rybicki BA, Sanders K, Peterson EL. Adult nutrient intake as a risk factor for Parkinson's disease. Int J Epidemiol 1999; 28(6): 1102-9.
[http://dx.doi.org/10.1093/ije/28.6.1102] [PMID: 10661654]

[46]   Fall PA, Fredrikson M, Axelson O, Granérus AK. Nutritional and occupational factors influencing the risk of Parkinson's disease: a case-control study in southeastern Sweden. Mov Disord 1999; 14(1): 28-37.
[http://dx.doi.org/10.1002/1531-8257(199901)14:1<28::AID-MDS1007>3.0.CO;2-O]   [PMID: 9918341]

[47]   Hellenbrand W, Boeing H, Robra BP, *et al.* Diet and Parkinson's disease. II: A possible role for the past intake of specific nutrients. Results from a self-administered food-frequency questionnaire in a case-control study. Neurology 1996; 47(3): 644-50.
[http://dx.doi.org/10.1212/WNL.47.3.644] [PMID: 8797457]

[48]   Alisky JM. Niacin improved rigidity and bradykinesia in a Parkinson's disease patient but also caused unacceptable nightmares and skin rash--a case report. Nutr Neurosci 2005; 8(5-6): 327-9.
[http://dx.doi.org/10.1080/10284150500484638] [PMID: 16669604]

[49]   Conn PF, Schalch W, Truscott TG. The singlet oxygen and carotenoid interaction. J Photochem Photobiol B 1991; 11(1): 41-7.
[http://dx.doi.org/10.1016/1011-1344(91)80266-K] [PMID: 1791493]

[50]   Di Mascio P, Kaiser S, Sies H. Lycopene as the most efficient biological carotenoid singlet oxygen quencher. Arch Biochem Biophys 1989; 274(2): 532-8.
[http://dx.doi.org/10.1016/0003-9861(89)90467-0] [PMID: 2802626]

[51]   Prema A, Janakiraman U, Manivasagam T, Thenmozhi AJ. Neuroprotective effect of lycopene against MPTP induced experimental Parkinson's disease in mice. Neurosci Lett 2015; 599: 12-9.
[http://dx.doi.org/10.1016/j.neulet.2015.05.024] [PMID: 25980996]

[52]   Kaur H, Chauhan S, Sandhir R. Protective effect of lycopene on oxidative stress and cognitive decline in rotenone induced model of Parkinson's disease. Neurochem Res 2011; 36(8): 1435-43.
[http://dx.doi.org/10.1007/s11064-011-0469-3] [PMID: 21484267]

[53]   Kostic AŽ, Milinˇci'c DD, Gaši'c UM, *et al.* Polyphenolicprofile and antioxidant properties of bee-collected pollen from sunflower (Helianthus annuusL) plant. LWT 2019; Vol. 112.

[54]   De-Melo AAM, Estevinho LM, Moreira MM, *et al.* Phenolic profile by HPLC-MS, biological potential, and nutritional value of apromising food: Monofloral bee pollen. J Food Biochem 2018; 42: 1-21.
[http://dx.doi.org/10.1111/jfbc.12536]

[55]   Yao LH, Jiang YM, Shi J, *et al.* Flavonoids in food and their health benefits. Plant Foods Hum Nutr 2004; 59(3): 113-22.
[http://dx.doi.org/10.1007/s11130-004-0049-7] [PMID: 15678717]

[56]   Kumar A, Sehgal N, Kumar P, Padi SSV, Naidu PS. Protective effect of quercetin against ICV colchicine-induced cognitive dysfunctions and oxidative damage in rats. Phytother Res 2008; 22(12): 1563-9.
[http://dx.doi.org/10.1002/ptr.2454] [PMID: 18980205]

[57]   Moretti M, Fraga DB, Rodrigues ALS. Preventive and therapeutic potential of ascorbic acid in neurodegenerative diseases. CNS Neurosci Ther 2017; 23(12): 921-9.

[http://dx.doi.org/10.1111/cns.12767] [PMID: 28980404]

[58]   Oudemans-van Straaten HM, Spoelstra-de Man AME, de Waard MC. Vitamin C revisited. Crit Care 2014; 18(4): 460.
[http://dx.doi.org/10.1186/s13054-014-0460-x] [PMID: 25185110]

[59]   Belluzzi E, Bisaglia M, Lazzarini E, Tabares LC, Beltramini M, Bubacco L. Human SOD2 modification by dopamine quinones affects enzymatic activity by promoting its aggregation: possible implications for Parkinson's disease. PLoS One 2012; 7(6): e38026.
[http://dx.doi.org/10.1371/journal.pone.0038026] [PMID: 22723845]

[60]   Milby K, Oke A, Adams RN. Detailed mapping of ascorbate distribution in rat brain. Neurosci Lett 1982; 28(1): 15-20.
[http://dx.doi.org/10.1016/0304-3940(82)90201-4] [PMID: 6121305]

[61]   Mefford IN, Oke AF, Adams RN. Regional distribution of ascorbate in human brain. Brain Res 1981; 212(1): 223-6.
[http://dx.doi.org/10.1016/0006-8993(81)90056-1] [PMID: 7225858]

[62]   Hansen SN, Tveden-Nyborg P, Lykkesfeldt J. Does vitamin C deficiency affect cognitive development and function? Nutrients 2014; 6(9): 3818-46.
[http://dx.doi.org/10.3390/nu6093818] [PMID: 25244370]

[63]   Hosoya K, Nakamura G, Akanuma S, Tomi M, Tachikawa M. Dehydroascorbic acid uptake and intracellular ascorbic acid accumulation in cultured Müller glial cells (TR-MUL). Neurochem Int 2008; 52(7): 1351-7.
[http://dx.doi.org/10.1016/j.neuint.2008.02.001] [PMID: 18353508]

[64]   Seitz G, Gebhardt S, Beck JF, *et al.* Ascorbic acid stimulates DOPA synthesis and tyrosine hydroxylase gene expression in the human neuroblastoma cell line SK-N-SH. Neurosci Lett 1998; 244(1): 33-6.
[http://dx.doi.org/10.1016/S0304-3940(98)00129-3] [PMID: 9578138]

[65]   Pardo B, Mena MA, Fahn S, García de Yébenes J. Ascorbic acid protects against levodopa-induced neurotoxicity on a catecholamine-rich human neuroblastoma cell line. Mov Disord 1993; 8(3): 278-84.
[http://dx.doi.org/10.1002/mds.870080305] [PMID: 8341291]

[66]   Sershen H, Reith MEA, Hashim A, Lajtha A. Protection against 1-methyl-4-phenyl-1,2-3,6-tetrahydropyridine neurotoxicity by the antioxidant ascorbic acid. Neuropharmacology 1985; 24(12): 1257-9.
[http://dx.doi.org/10.1016/0028-3908(85)90163-7] [PMID: 3879338]

[67]   Nagayama H, Hamamoto M, Ueda M, Nito C, Yamaguchi H, Katayama Y. The effect of ascorbic acid on the pharmacokinetics of levodopa in elderly patients with Parkinson disease. Clin Neuropharmacol 2004; 27(6): 270-3.
[http://dx.doi.org/10.1097/01.wnf.0000150865.21759.bc] [PMID: 15613930]

[68]   Fox SH, Katzenschlager R, Lim SY, *et al.* Movement Disorder Society Evidence-Based Medicine Committee. International Parkinson and movement disorder society evidence-based medicine review: Update on treatments for the motor symptoms of Parkinson's disease. Mov Disord 2018; 33(8): 1248-66.
[http://dx.doi.org/10.1002/mds.27372] [PMID: 29570866]

[69]   He XB, Kim M, Kim SY, *et al.* Vitamin C facilitates dopamine neuron differentiation in fetal midbrain through TET1- and JMJD3-dependent epigenetic control manner. Stem Cells 2015; 33(4): 1320-32.
[http://dx.doi.org/10.1002/stem.1932] [PMID: 25535150]

[70]   Yan J, Studer L, McKay RDG. Ascorbic acid increases the yield of dopaminergic neurons derived from basic fibroblast growth factor expanded mesencephalic precursors. J Neurochem 2001; 76(1): 307-11.
[http://dx.doi.org/10.1046/j.1471-4159.2001.00073.x] [PMID: 11146004]

[71]   Lee JY, Chang MY, Park CH, *et al.* Ascorbate-induced differentiation of embryonic cortical precursors into neurons and astrocytes. J Neurosci Res 2003; 73(2): 156-65.
[http://dx.doi.org/10.1002/jnr.10647] [PMID: 12836158]

[72]   Wulansari N, Kim EH, Sulistio YA, Rhee YH, Song JJ, Lee SH. Vitamin C-induced epigenetic modifications in donor NSCs establish midbrain marker expressions critical for cell-based therapy in Parkinson's disease. Stem Cell Reports 2017; 9(4): 1192-206.
[http://dx.doi.org/10.1016/j.stemcr.2017.08.017] [PMID: 28943252]

[73]   Hughes KC, Gao X, Kim IY, *et al.* Intake of antioxidant vitamins and risk of Parkinson's disease. Mov Disord 2016; 31(12): 1909-14.
[http://dx.doi.org/10.1002/mds.26819] [PMID: 27787934]

[74]   Yang F, Wolk A, Håkansson N, Pedersen NL, Wirdefeldt K. Dietary antioxidants and risk of Parkinson's disease in two population-based cohorts. Mov Disord 2017; 32(11): 1631-6.
[http://dx.doi.org/10.1002/mds.27120] [PMID: 28881039]

[75]   Miyake Y, Fukushima W, Tanaka K, *et al.* Fukuoka Kinki Parkinson's Disease Study Group. Dietary intake of antioxidant vitamins and risk of Parkinson's disease: a case-control study in Japan. Eur J Neurol 2011; 18(1): 106-13.
[http://dx.doi.org/10.1111/j.1468-1331.2010.03088.x] [PMID: 20491891]

[76]   Cherubini AMartin, Andres-Lacueva C, *et al.* Vitamin E levels, cognitive impairment and dementia in older persons: the InCHIANTI study. Neurobiology of Aging 2005; 26(7): 987-94.

[77]   Cesari M, Pahor M, Bartali B, *et al.* Antioxidants and physical performance in elderly persons: the Invecchiare in Chianti (InCHIANTI) study. Am J Clin Nutr 2004; 79(2): 289-94.
[http://dx.doi.org/10.1093/ajcn/79.2.289] [PMID: 14749236]

[78]   Beharka A, Redican S, Leka L, Meydani SN. Vitamin E status and immune function. Methods Enzymol 1997; 282: 247-63.
[http://dx.doi.org/10.1016/S0076-6879(97)82112-X] [PMID: 9330293]

[79]   Clarke MW, Burnett JR, Croft KD. Vitamin E in human health and disease. Crit Rev Clin Lab Sci 2008; 45(5): 417-50.
[http://dx.doi.org/10.1080/10408360802118625] [PMID: 18712629]

[80]   Aparicio JM, Bélanger-Quintana A, Suárez L, *et al.* Ataxia with isolated vitamin E deficiency: case report and review of the literature. J Pediatr Gastroenterol Nutr 2001; 33(2): 206-10.
[http://dx.doi.org/10.1097/00005176-200108000-00022] [PMID: 11568526]

[81]   Cadet JL, Katz M, Jackson-Lewis V, Fahn S. Vitamin E attenuates the toxic effects of intrastriatal injection of 6-hydroxydopamine (6-OHDA) in rats: behavioral and biochemical evidence. Brain Res 1989; 476(1): 10-5.
[http://dx.doi.org/10.1016/0006-8993(89)91530-8] [PMID: 2492442]

[82]   Sharma N, Nehru B. Beneficial effect of vitamin E in rotenone induced model of PD: behavioural, neurochemical and biochemical study. Exp Neurobiol 2013; 22(3): 214-23.
[http://dx.doi.org/10.5607/en.2013.22.3.214] [PMID: 24167416]

[83]   Roghani M, Behzadi G. Neuroprotective effect of vitamin E on the early model of Parkinson's disease in rat: behavioral and histochemical evidence. Brain Res 2001; 892(1): 211-7.
[http://dx.doi.org/10.1016/S0006-8993(00)03296-0] [PMID: 11172767]

[84]   Heim C, Kolasiewicz W, Kurz T, Sontag KH. Behavioral alterations after unilateral 6-hydroxydopamine lesions of the striatum. Effect of alpha-tocopherol. Pol J Pharmacol 2001; 53(5): 435-48.
[PMID: 11990061]

[85]   Perry TL, Yong VW, Hansen S, *et al.* α-tocopherol and β-carotene do not protect marmosets against the dopaminergic neurotoxicity of N-methyl-4-phenyl-1,2,3,6-tetrahydropyridine. J Neurol Sci 1987; 81(2-3): 321-31.

[http://dx.doi.org/10.1016/0022-510X(87)90106-7] [PMID: 3121800]

[86] Comitato R, Nesaretnam K, Leoni G, *et al.* A novel mechanism of natural vitamin E tocotrienol activity: involvement of ERbeta signal transduction. Am J Physiol Endocrinol Metab 2009; 297(2): E427-37.
[http://dx.doi.org/10.1152/ajpendo.00187.2009] [PMID: 19491296]

[87] Nakaso K, Tajima N, Horikoshi Y, *et al.* The estrogen receptor β-PI3K/Akt pathway mediates the cytoprotective effects of tocotrienol in a cellular Parkinson's disease model. Biochim Biophys Acta 2014; 1842(9): 1303-12.
[http://dx.doi.org/10.1016/j.bbadis.2014.04.008] [PMID: 24768803]

[88] de Rijk MC, Breteler MM, den Breeijen JH, *et al.* Dietary antioxidants and Parkinson disease. The Rotterdam Study. Arch Neurol 1997; 54(6): 762-5.
[http://dx.doi.org/10.1001/archneur.1997.00550180070015] [PMID: 9193212]

[89] Fahn S. A pilot trial of high-dose alpha-tocopherol and ascorbate in early Parkinson's disease. Ann Neurol 1992; 32(S1) (Suppl.): S128-32.
[http://dx.doi.org/10.1002/ana.410320722] [PMID: 1510371]

[90] Vatassery GT, Fahn S, Kuskowski MA. Parkinson Study Group. Alpha tocopherol in CSF of subjects taking high-dose vitamin E in the DATATOP study. Neurology 1998; 50(6): 1900-2.
[http://dx.doi.org/10.1212/WNL.50.6.1900] [PMID: 9633757]

[91] Molina JA, de Bustos F, Jiménez-Jiménez FJ, *et al.* Cerebrospinal fluid levels of alpha-tocopherol (vitamin E) in Parkinson's disease. J Neural Transm (Vienna) 1997; 104(11-12): 1287-93.
[http://dx.doi.org/10.1007/BF01294729] [PMID: 9503274]

[92] Férnandez-Calle P, Molina JA, Jiménez-Jiménez FJ, *et al.* Serum levels of alpha-tocopherol (vitamin E) in Parkinson's disease. Neurology 1992; 42(5): 1064-6.
[http://dx.doi.org/10.1212/WNL.42.5.1064] [PMID: 1579230]

[93] Dexter DT, Ward RJ, Wells FR, *et al.* α-tocopherol levels in brain are not altered in Parkinson's disease. Ann Neurol 1992; 32(4): 591-3.
[http://dx.doi.org/10.1002/ana.410320420] [PMID: 1456747]

[94] Logroscino G, Marder K, Cote L, Tang MX, Shea S, Mayeux R. Dietary lipids and antioxidants in Parkinson's disease: a population-based, case-control study. Ann Neurol 1996; 39(1): 89-94.
[http://dx.doi.org/10.1002/ana.410390113] [PMID: 8572672]

[95] Eyles DW, Burne THJ, McGrath JJ, Vitamin D. Vitamin D, effects on brain development, adult brain function and the links between low levels of vitamin D and neuropsychiatric disease. Front Neuroendocrinol 2013; 34(1): 47-64.
[http://dx.doi.org/10.1016/j.yfrne.2012.07.001] [PMID: 22796576]

[96] Sanchez B, Lopez-Martin E, Segura C, Labandeira-Garcia JL, Perez-Fernandez R. 1,25-Dihydroxyvitamin D(3) increases striatal GDNF mRNA and protein expression in adult rats. Brain Res Mol Brain Res 2002; 108(1-2): 143-6.
[http://dx.doi.org/10.1016/S0169-328X(02)00545-4] [PMID: 12480187]

[97] Christakos S, Dhawan P, Verstuyf A, Verlinden L, Carmeliet G. Vitamin D: metabolism, molecular mechanism of action, and pleiotropic effects. Physiol Rev 2016; 96(1): 365-408.
[http://dx.doi.org/10.1152/physrev.00014.2015] [PMID: 26681795]

[98] Kulda V. Metabolizmus vitaminu D [Vitamin D metabolism]. Vnitr Lek. 2012 May;58(5):400-4. Czech.
[PMID: 22716179]

[99] Carlberg C, Molnár F. Vitamin D receptor signaling and its therapeutic implications: Genome-wide and structural view. Can J Physiol Pharmacol 2015; 93(5): 311-8.
[http://dx.doi.org/10.1139/cjpp-2014-0383] [PMID: 25741777]

[100] Haussler MR, Whitfield GK, Kaneko I, *et al.* Molecular mechanisms of vitamin D action. Calcif

Tissue Int 2013; 92(2): 77-98.
[http://dx.doi.org/10.1007/s00223-012-9619-0] [PMID: 22782502]

[101]   Kono K, Fujii H, Nakai K, *et al*. Anti-oxidative effect of vitamin D analog on incipient vascular lesion in non-obese type 2 diabetic rats. Am J Nephrol 2013; 37(2): 167-74.
[http://dx.doi.org/10.1159/000346808] [PMID: 23406697]

[102]   Myszka M, Klinger M. [The immunomodulatory role of Vitamin D]. Postepy Hig Med Dosw 2014; 68: 865-78.
[http://dx.doi.org/10.5604/17322693.1110168] [PMID: 24988607]

[103]   Samuel S, Sitrin MD. Vitamin D's role in cell proliferation and differentiation. Nutr Rev 2008; 66(10) (Suppl. 2): S116-24.
[http://dx.doi.org/10.1111/j.1753-4887.2008.00094.x] [PMID: 18844838]

[104]   Olajide OJ, Yawson EO, Gbadamosi IT, *et al*. Ascorbic acid ameliorates behavioural deficits and neuropathological alterations in rat model of Alzheimer's disease. Environ ToxicolPharmacol 2017; 50: 200-11.

[105]   Yoon JH, Park DK, Yong SW, Hong JM. Vitamin D deficiency and its relationship with endothelial dysfunction in patients with early Parkinson's disease. J Neural Transm (Vienna) 2015; 122(12): 1685-91.
[http://dx.doi.org/10.1007/s00702-015-1452-y] [PMID: 26343034]

[106]   Sahota O. Understanding vitamin D deficiency. Age Ageing 2014; 43(5): 589-91.
[http://dx.doi.org/10.1093/ageing/afu104] [PMID: 25074537]

[107]   Köstner K, Denzer N, Müller CS, Klein R, Tilgen W, Reichrath J. The relevance of vitamin D receptor (VDR) gene polymorphisms for cancer: a review of the literature. Anticancer Res 2009; 29(9): 3511-36.
[PMID: 19667145]

[108]   Burne THJ, McGrath JJ, Eyles DW, Mackay-Sim A. Behavioural characterization of vitamin D receptor knockout mice. Behav Brain Res 2005; 157(2): 299-308.
[http://dx.doi.org/10.1016/j.bbr.2004.07.008] [PMID: 15639181]

[109]   Suzuki M, Yoshioka M, Hashimoto M, *et al*. 25-hydroxyvitamin D, vitamin D receptor gene polymorphisms, and severity of Parkinson's disease. Mov Disord 2012; 27(2): 264-71.
[http://dx.doi.org/10.1002/mds.24016] [PMID: 22213340]

[110]   Gatto NM, Paul KC, Sinsheimer JS, *et al*. Vitamin D receptor gene polymorphisms and cognitive decline in Parkinson's disease. J Neurol Sci 2016; 370: 100-6.
[http://dx.doi.org/10.1016/j.jns.2016.09.013] [PMID: 27772736]

[111]   Li C, Qi H, Wei S, *et al*. Vitamin D receptor gene polymorphisms and the risk of Parkinson's disease. Neurol Sci 2015; 36(2): 247-55.
[http://dx.doi.org/10.1007/s10072-014-1928-9] [PMID: 25169913]

[112]   Campos FL, Cristovão AC, Rocha SM, Fonseca CP, Baltazar G. GDNF contributes to oestrogen-mediated protection of midbrain dopaminergic neurones. J Neuroendocrinol 2012; 24(11): 1386-97.
[http://dx.doi.org/10.1111/j.1365-2826.2012.02348.x] [PMID: 22672424]

[113]   Sanchez B, Relova JL, Gallego R, Ben-Batalla I, Perez-Fernandez R. 1,25-Dihydroxyvitamin D3 administration to 6-hydroxydopamine-lesioned rats increases glial cell line-derived neurotrophic factor and partially restores tyrosine hydroxylase expression in substantia nigra and striatum. J Neurosci Res 2009; 87(3): 723-32.
[http://dx.doi.org/10.1002/jnr.21878] [PMID: 18816795]

[114]   Kim JS, Ryu SY, Yun I, *et al*. 1α,25-Dihydroxyvitamin D3 protects dopaminergic neurons in rodent models of Parkinson's disease through inhibition of microglial activation. J Clin Neurol 2006; 2(4): 252-7.
[http://dx.doi.org/10.3988/jcn.2006.2.4.252] [PMID: 20396528]

[115] Sanchez B, Lopez-Martin E, Segura C, Labandeira-Garcia JL, Perez-Fernandez R. 1,25-Dihydroxyvitamin D(3) increases striatal GDNF mRNA and protein expression in adult rats. Brain Res Mol Brain Res 2002; 108(1-2): 143-6.
[http://dx.doi.org/10.1016/S0169-328X(02)00545-4] [PMID: 12480187]

[116] Jang W, Kim HJ, Li H, *et al.* 1,25-Dyhydroxyvitamin D_3 attenuates rotenone-induced neurotoxicity in SH-SY5Y cells through induction of autophagy. Biochem Biophys Res Commun 2014; 451(1): 142-7.
[http://dx.doi.org/10.1016/j.bbrc.2014.07.081] [PMID: 25078626]

[117] Knekt P, Kilkkinen A, Rissanen H, Marniemi J, Sääksjärvi K, Heliövaara M. Serum vitamin D and the risk of Parkinson disease. Arch Neurol 2010; 67(7): 808-11.
[http://dx.doi.org/10.1001/archneurol.2010.120] [PMID: 20625085]

[118] Evatt ML, Delong MR, Khazai N, Rosen A, Triche S, Tangpricha V. Prevalence of vitamin d insufficiency in patients with Parkinson disease and Alzheimer disease. Arch Neurol 2008; 65(10): 1348-52.
[http://dx.doi.org/10.1001/archneur.65.10.1348] [PMID: 18852350]

[119] Kenborg L, Lassen CF, Ritz B, *et al.* Outdoor work and risk for Parkinson's disease: a population-based case-control study. Occup Environ Med 2011; 68(4): 273-8.
[http://dx.doi.org/10.1136/oem.2010.057448] [PMID: 20884793]

[120] Kravietz A, Kab S, Wald L, *et al.* Association of UV radiation with Parkinson disease incidence: A nationwide French ecologic study. Environ Res 2017; 154: 50-6.
[http://dx.doi.org/10.1016/j.envres.2016.12.008] [PMID: 28033496]

[121] Wang J, Yang D, Yu Y, Shao G, Wang Q. Vitamin D and sunlight exposure in newly-diagnosed Parkinson's disease. Nutrients 2016; 8(3): 142.
[http://dx.doi.org/10.3390/nu8030142] [PMID: 26959053]

CHAPTER 5

Potential of Gut Microbiome in the Diagnosis and Treatment of Alzheimer's and Parkinson's Disease

Nilofar Khan[1] and **Ravishankar Patil**[1,*]

Amity Institute of Biotechnology, Amity University, Maharashtra 410206, India.

Abstract: Neurodegenerative diseases (NDD) are a heterogeneous group of disorders characterized by a progressive, selective loss of physiologically related neuronal systems. Some prominent diseases include Alzheimer's disease (AD), Parkinson's disease (PD), amyotrophic lateral sclerosis (ALS), Multiple Sclerosis (MS), and Huntington's disease (HD). It is believed that oxidative stress-induced cellular degeneration, inflammation, mitochondrial involvement, and dysfunction are important aspects in the pathogenesis of NDDs. Despite many decades of research and intensive studies, it has been an unending struggle to discover the root cause and a cure for these life-threatening ailments. However, the emerging domains of research provide evidence that probiotics and human gut microflora have a peculiar relationship with health and the pathogenesis of several diseases, including NDDs.

Microbiome and nutrients have a profound impact on the brain by influencing their development and function in health and diseases. The gut ecosystem and any modulation thereof exhibit a significant impact on the physiological and psychological health of an individual. The present chapter discusses the effect of the beneficial gut microbial community *versus* pathogens on the overall human health and its role in the development, diagnosis, and management of NDDs, especially Alzheimer's disease (AD) and Parkinson's disease (PD). Furthermore, the potency of probiotics and prebiotics as a gut-friendly therapeutic agent to treat these disorders is highlighted.

Keywords: Alzheimer's disease, Antioxidant, Gut microbiome, Gut-brain axis, Neurodegenerative diseases, Parkinson's disease, Prebiotics, Probiotics.

INTRODUCTION

Neurodegenerative diseases are ailments associated with the central nervous system (CNS), identified as chronic and progressive, and are indicated by the loss of neurons from specific areas of the brain [1]. The most communal NDDs are Alzheimer's disease (AD) and Parkinson's disease (PD). The progressive loss of neuronal cells and synapses, resulting in memory loss, poor learning, and

---

* **Corresponding author Ravishankar Patil:** Amity Institute of Biotechnology, Amity University, Maharashtra, 410206, India; Tel: +91 8999754168; E-mail: ravishankarpatil1@gmail.com

Sachchida Nand Rai (Ed.)

cognition, are the major characteristics of AD [2]. The depletion of acetylcholine level is the main molecular symptom of AD. Two critical hypotheses of AD pathogenesis is amyloid-beta (Aβ) [3] and tau [4]. Moreover, lifestyle factors such as being sedentary, stress, poor diets, and tobacco and alcohol consumption are also contributing factors in AD development [2, 5]. PD is considered the second most common NDD after AD. The confirmatory sign of PD is the depletion of dopaminergic neurons present in the substantia nigra region of the brain. The primary symptoms of PD are related to movement and balance (motor symptoms); however, secondary symptoms are also found, such as depression, dementia, hyposmia, chronic fatigue, constipation, loss of smell sense, and sleep disturbance [6].

Though numbers of drugs are in practice for the management of either AD or PD, no single one has proved to be effective for their complete cure. Hence there is a necessity for novel potent, and safer therapies that could offer symptomatic relief along with prevention of disease pathogenesis and progression. Recent research suggests that different disorders of the central nervous system (CNS) have a strong connection with gut microbiota *via* the enteric nervous system (ENS). Scientific literature proved that, in the diseased subject, there is a tremendous change in gut microbiota, which might be accountable for the increasing severity of illness. A decrease in gut-friendly microbes (For example, *Lactobacilli, Bacteroides, Prevotella, Bifidobacterium, etc.*) and an increase in pathogenic microbial population (For example, *Enterobacteria, Streptococci, Staphylococci, Shigella, H. pylori, etc.*) in case of AD and PD have been studied in detailed [7 - 13]. Gut microbiota affects brain functioning through the Gut-Brain Axis (GBA) in normal as well as diseased conditions. The microbial dysbiosis alters gut permeability, increases chronic inflammation, and triggers AD development [14 - 16]. A number of deadly pathogens, including *Staphylococcus aureus,* and *Mycobacterium tuberculosis,* synthesize amyloid protein, a protein having a key role in plaque formation in AD [17, 18]. In PD, gut microbes affect dopamine synthesis, α-synuclein deposition, increases oxidative stress, local inflammation, intestinal permeability, and causes constipation [19]. Current research suggests that dysfunction of gut microbiota can be explored for the early diagnosis of PD [19], and precise modulation of gut microflora in the diseased condition could help for effective management of disease pathogenesis as well as progression.

The present chapter discusses pathogenesis, prevalence, symptoms, and current therapies of AD and PD (in brief). More efforts have been taken to review the literature on the interaction of gut microbiota with CNS and its further involvement in the pathological process of AD and PD. Also, the potential of probiotics and prebiotics as therapeutic agents is discussed in deriving possible treatment strategies to combat these NDDs by balancing the gut microbiome ecosystem.

ALZHEIMER'S DISEASE: PATHOMECHANISM, PREVALENCE, SYMPTOMS, AND TREATMENT

Alzheimer's disease (AD) is an NDD, causing a chronic and exponential loss of function and neuronal degeneration along with psychological distress [20]. It is associated with tau and amyloid peptide deposition in parts of the brain, which causes neurons to lose their function, mostly affecting neocortical structures [21]. The hallmark of AD is neuritic plaques and neurofibrillary tangles pertaining to amyloid-beta peptides (Aβ) accumulation in tissues of the brain and cytoskeleton changes caused by the hyperphosphorylation of tau protein present in the neurons [22]. The cognitive symptoms of AD are short-term memory, praxis, and executive and visuospatial dysfunction [23]. There are various beta-amyloid isoforms that vary depending on the amino acids present on the C-terminal. However, $A\beta1-42$ peptides play a vital role in AD pathogenesis [24]. The age, genetic constructs, and environmental factors impart a shift in metabolism that catalyzes the progression of amyloidogenesis of APP (Amyloid precursor protein) in deteriorating the physiological pathways, thereby promoting the APP cleavage using BACE-1 (beta-secretase enzyme) [25]. This particular reaction is the prime backbone of Aβ production [25]. The neurotoxic potential exuded by the Aβ peptides aid their aggregation into oligomers (insoluble) and protofibrils. Additionally, depleting Aβ from the brain causes their accumulation outside neurons, thereby triggering cascades leading to cytoskeletal modification, neural dysfunction, and finally, apoptosis [26].

Recent studies propose that AD may be categorized into three clinical stages: (i) pre-clinical symptoms of AD, last for many decades until Aβ accumulation and their excessive production; (ii) early-stage pathology is the pre-dementia phase; (iii) Aggregation of neuritic plaques along with neurofibrillary tangles in certain areas of the brain, clinically defined stage of AD is assumed [27].

Prevalence of AD

Millions of people are affecting by AD, and notably, it is the most dominant cause of dementia (around 60-80%). However, delaying the onset of symptoms by a year significantly reduces the AD prevalence by more than 9 million cases over the next 4 decades [28]. Autopsies suggested that the neuropathological changes observed in patients with AD in developed countries are similar to the qualitative changes observed in developed countries [29].

Symptoms of AD

As earlier discussed, AD is known to have a subjective impact on patients, with a symptom being the inability to remember and form new memories. This is due to the disruption of neurons *via* amyloid plaque deposition [30]. The following are some prominent symptoms of AD [31 - 33]:

- Difficulty in the ability to solve problems
- Inefficiency in doing personal care and work
- Loss of memory affecting the patient's day to day life
- Inability to process visual images
- Poor eye coordination and judgment
- Psychological illnesses

Causes of AD

Various studies have suggested that multiple factors can aggravate the onset of AD and dementia. The most common causes include vascular risk factors such as psychological illnesses, diabetes, hypertension and obesity, smoking, and less mental or physical activity [34]. The possibility of having AD increases if an individual has a history of NDD [35 - 37]. People having more than one immediate relative with AD are at higher risk. This is because, along with environmental factors and lifestyle, hereditary and genetics also play a significant role in the development of AD [38]. Consistent mental stimulation has been known to lower the risk of AD by building memory [39]. Many post-mortem studies in AD patients suggested that a significant number of cholinergic abnormalities like decreased acetylcholine release, loss of Acetylcholine receptor (AChR) expression early in the course of disease lead to cognitive and non-cognitive behavioral decline along with the deposition of neuronal plaque [40].

AChR bind to neurotransmitter- Acetylcholine (Ach) in the CNS and PNS associated with selective thinking, processing, and sensory inputs [41, 42]. AChR is vital as they are abundantly expressed throughout the mammalian brain and regulate learning, memory, sensory reception, neuroprotection, and memory formation [43]. Studies have reflected that episodic memory impairment seen in AD corresponds to disrupted cholinergic neurotransmission, indicating abnormal AChR function [44]. However, uncertainty looms over the pharmacological effects of Aβ on a subtype AChR- the α7 AChR [45]. It is suggested that modifications of the α7 AChR could deem to be an effective treatment of AD

[46]. Studies have hypothesized that accumulation of Aβ in AD blocks normal AChR (nAChR) activation, given its inhibitory effect.

Treatment Strategies for AD

Certain enzymes and pathways have been speculated to possibly bring relief from distress caused by this disease. However, unfortunately, no drug is yet developed for the complete management of AD.

Acetylcholine Esterase (Ache)

The enzyme acetylcholinesterase (AChE) plays a major role in ACh degradation, which is vital in AD [47]. AChE inhibitors are being effectively used in the AD treatment to balance the level of Ach [48].

β Secretase

It is the enzyme associated with cleavage of APP at β-site; BACE [49]. β secretase has been considered as a significant treatment compound due to its ability to catalyze the initiation of Aβ production [50].

ϒ Secretase

This enzyme determines the proportion of Aβ $_{1-40}$ to Aβ $_{1-42}$, along with the ability to cleave substrates [49, 51].

Aβ Vaccination

The anti-Aβ monoclonal antibodies (mAbs) are specific in terms of their epitopes and target Aβ peptides [52]. These vaccines possess anti-aggregation properties *in vitro* and target the N-terminus of Aβ, dissolving monomers, oligomers, and fibrils. The sequestration and binding of soluble Aβ may be linked to improvement in memory, which is promoted by passive immunisation [53].

Amyloid Anti-Aggregation Therapies

Synthetic peptides have been designed by researchers that are similar to the active site region of Aβ and can destabilize the β sheet structures potentially [54]. These anti-aggregates are under clinical trials, which could be used as a potential therapy.

Anti-Inflammatory Drugs

Reduced margin of susceptibility to AD was reported with the chronic administration of nonsteroidal anti-inflammatory drugs [55] that have shown reduced inflammation in the plaques [56].

Cholesterol-Lowering Drugs

Cholesterol depletion in the nerve cells significantly reduced the Aβ secretion [57, 58]. With statins being administered chronically, the risk of AD was reduced in individuals [59]. However, detailed research is required to clarify the role of cholesterol in APP processing [60].

PARKINSON'S DISEASE: PATHOMECHANISM, PREVALENCE, SYMPTOMS, AND TREATMENT

Parkinson's disease (PD) is the second most prevalent NDD after AD, which is distinguished by the deterioration of dopaminergic neurons progressively by forming filamentous inclusion components called Lewy bodies (LB) within the nerve cells [61]. Mutation in some genes like α-synuclein relates to PD through its α-synuclein accumulation as LB and/or neuritis [62, 63]. The sporadic occurrence of PD encompasses about 90% of the cases. Irrespective of several decades' research efforts, the precise pathogenesis of PD remains unknown; however, oxidative stress, protein misfolding, and mitochondrial dysfunction have a pivotal role in the pathology of this ailment [64]. In sporadically occurring PD, some non-genetic environmental factors/toxins act as triggers [65, 66]. Also, smoking and older age are the dominant risk factors for sporadic PD apart from alcoholism, coffee consumption, and exposure to hazardous chemicals [67]. Unfortunately, primary PD symptoms are seen after the degeneration of 60-80% dopaminergic neurons [68], making the management of the disease difficult.

Prevalence of PD

PD is known as an idiopathic disorder and it may occur sporadically, showing no identifiable cause. However, it is inherited as both autosomal dominant and recessive traits in some families [69]. From 1990 to 2015, the total count of diseased individuals globally has grown more than double [70]. The incidence and disease span define the prevalence of disease while being associated with protective and risk factors [71, 72].

Symptoms of PD

The characteristic motor symptoms (primary) of PD are rigidity, tremors, postural instability, bradykinesia, and parkinsonian gait [73, 74, 75]. However, secondary symptoms such as depression, dementia, hyposmia, chronic fatigue, constipation, loss of smell sense, and sleep disturbance [6] have also been found as the disease worsens.

Treatment Strategies for PD

Unfortunately, there have been no treatments available to cure PD. However, the therapies at present can only delay the PD symptoms. Currently, the PD treatments are aimed at balancing the optimum level of dopamine either by inhibition of its further metabolism or by using its precursor [76].

Dopamine Replacement Therapy (DRT)

DRT aids in increasing dopamine and is employed as a standard treatment by regulating and controlling different motor symptoms of PD. This therapy is used to elevate the dopamine levels in the brain by either i) replacing or ii) mimicking dopamine by inhibiting its breakdown [77]. It includes DA precursor (L-DOPA), Peripheral decarboxylase inhibitors (Benzserazide and Carbidopa), Dopaminergic agonists (Apomorphine, bromocriptine, *etc.*), MAO inhibitor (Selegiline, Rasagiline, *etc.*), Anticholinergic drugs (Benzhexol and orphenadrine), Dopamine Facilitators (Amantadine), Catechol-O Methyl Transferase (COMT) Inhibitors (Tolcapone and Entacapone) and dopamine reuptake inhibitor (Amphetamine) [78, 79, 80]

L-DOPA: Promising DRT for PD

L-DOPA (L 3, 4 dihydroxy phenylalanine) is a non-protein amino acid that is the prime gold standard drug used for PD treatment, which gives symptomatic relief [81, 82]. Naturally, it is found in the plants of the genus *Mucuna* and since ancient times being used as a promising treatment for PD [83, 84, 85, 86, 87]. In animals, it occurs in the medulla of the adrenal gland and other nervous tissues. L-DOPA is further decarboxylated to become dopamine [88]. It is essentially used as a clinical treatment in PD and dopamine responsive dystonia [89].

Anti-inflammatory Agents and Protein Anti-Aggregants

This strategy has been developed to protect neurons from apoptosis. Inflammatory reactions like microglial activation are distinctly present in large numbers in the brains of PD patients. Thus, anti-inflammatory agents may prove to be effective in protecting the neuronal cells during the progression of PD [90]. Hence, protein anti-aggregants may be a promising approach to PD treatment.

Advanced treatment of PD, such as Deep Brain Stimulation (DBS), is also a reliable and established surgical treatment for PD. The use of DBS of the subthalamic nucleus (STN) is an advanced treatment of PD in which a medical device called a pacemaker is employed [91].

ROLE OF GUT MICROBIOME IN METABOLISM AND HEALTH

The gut microbiome is the ecosystem of microorganisms such as bacteria, archaebacteria, fungi, *etc.*, in the gut of a host (human), which may exude positive impact and improve overall health [92]. These microbes govern various fundamental functions in humans like metabolism of essential substances, production of short-chain fatty acids (SCFA), fermenting undigested carbohydrates, synthesis of vitamin B and K, and protection from pathogens [92]. The genesis of the microbiome in the gut is an exceptional event that opens many portals for the researchers to explore in time. A variety of factors influence the constitution and quality of microbiota residing in the human gut, like their age, diet, exposure to antibiotics, mode of delivery, sanitation [93, 94]. The human gut microbiome can be classified as opportunistic (causative of infections) or beneficial (called the 'cleansers' of the gut). *Lactobacterium* species *Bifidobacterium (B. bifidum)*, (*L. rhamnosus, L. Plantarum, L. acidophilus, Peptostreptococci, Propionobacteria,* and *Enterococci,* belong to the beneficial group exerting many benefits to our body. Whereas, *Enterobacteria, Actinobacteria, Clostridia, Staphylococci, Peptococci, Bacteriodes, Bacilli, Streptococci,* and Yeasts, *etc.* comprise an opportunistic group of microbes [95] inducing a negative impact to the body *via* some toxic metabolic activities. The imbalance in the population of the microbes would lead to a condition of dysbiosis that causes GI distress like Irritable Bowel Syndrome (IBS) and dysfunctional metabolism [96, 97].

Gut dysbiosis has a strong connection with the pathogenesis of different diseases and can be counteracted using probiotics. Probiotics are live microorganisms exerting beneficial effects on the health and physiology of the host. Therefore, the introduction of beneficial bacterial species to the GI tract may be a very attractive option to re-establish the microbial equilibrium and prevention of diseases [98]. A potential microbe can be screened for a set of standardized parameters that consider an organism probiotic.

The selection criteria may include characteristics [99] such as:

1. Immune to the effects of bile salts and pancreatic enzymes
2. Resistant toward gastric acidity
3. Adherence to intestinal mucosa cells
4. Should live through transporting and storing for improved adherence
5. Lacking translocation and synthesis of antimicrobial substances

Role of Prebiotics in Gut Microflora Balance

With time and growing research, it is seen that diet has played a vital role in conditioning susceptibility and acting as a risk factor for NDDs. Conversely, it has also gained recognition as a potential therapeutic strategy along with the treatments and medication. Prebiotics are fiber enriched compounds that boost the growth of essential microorganisms in the gut. Certain pieces of evidence are showing a particular diet is proving helpful for the PD *via* the positive metabolic activity of gut microflora and suppression of harmful microbial community actions [100, 101].

Gut Microbiome and Neurodegenerative Disease

The gut has shown to have close communication with the nervous system. The components of the gut interact with the parasympathetic and sympathetic nervous system *via* the Gut-Brain Axis [16]. It is derived that the CNS relays communication through autonomic pathways (ANS) from the mucosal layer and muscle lining of the gut. By this, the brain manages and promotes mobility, permeability, immunity, and secretion of mucus [102, 103, 104]. The disturbance in the gut microbial environment in the early onset of AD reflects a change in the behavioral attributes, especially on the psychiatric health of the individual like anxiety and depression, chronic pain, MS, PD, and AD [105]. It regulates the bidirectional communication between the gut and brain, indicating that changes in gut microbiota can influence the behavioral, physiological, and cognitive functions of the brain [106, 107]. It is inferred that neurotransmitters and neuromodulators can be synthesized and released by gut microflora like dopamine and histamine, serotonin, glutamate, biogenic amines, short-chain fatty acids, metabolites like GABA, tryptophan, and homocysteine [108]. Variable suited combinations of specific nutrients, inclusive of neuronal precursors and cofactors, can prevent synaptic loss and may decelerate membrane- related pathogenesis in ENS and CNS [109]. Several reports strongly suggest that gut microflora extensively synthesizes active neuronal molecules that may regulate the NDD pathogenesis like PD and AD [13, 110, 111, 112, 113]. Dysbiosis mediated pathogenesis of AD and PD with possible treatment using probiotics and prebiotics is represented in Fig. (**1**).

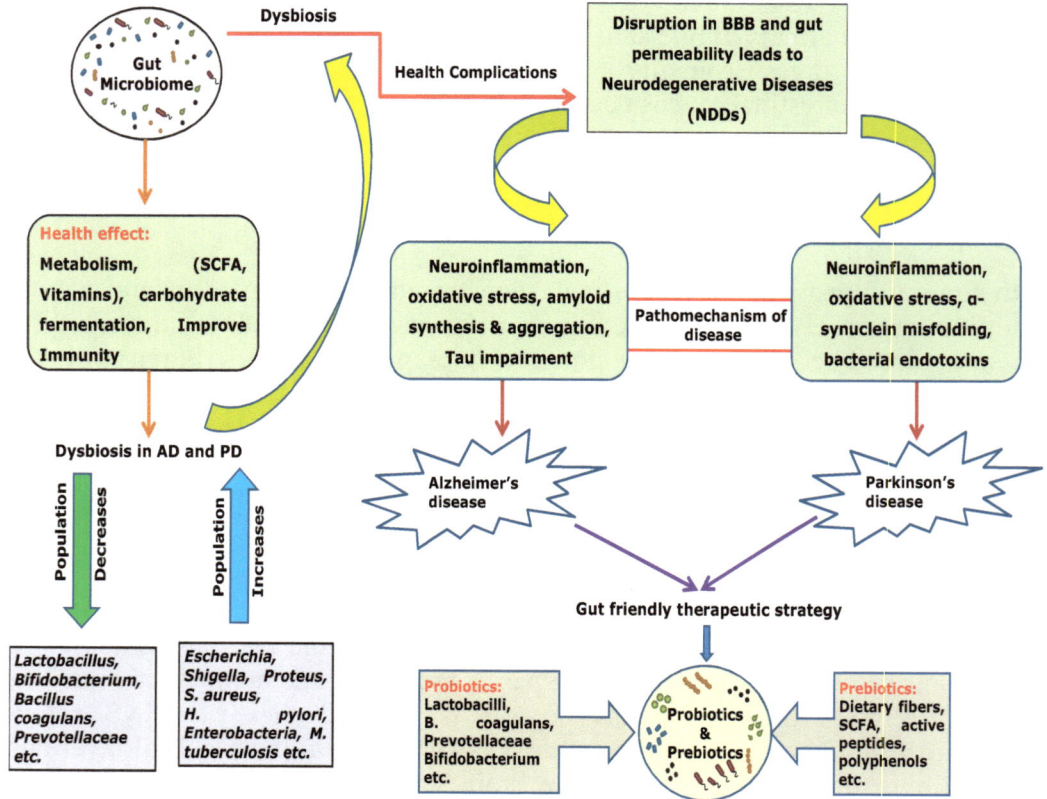

Fig. (1). Role of the gut microbiome in health and diseases generation and possible treatment using probiotics and prebiotics.

Gut and Alzheimer's Disease

In early neuropathology, AD was believed to be an NDD exclusively without any interaction with other parts of the body. However, this notion changed when it was discovered that the blood-brain barrier (BBB) and GI tract played a vital role as the protectors of the nervous system and possibly dominated the AD pathogenesis [114]. It is observed that inflammatory reactions and amyloid aggregation in elderly individuals being triggered by the permeability of the gut and BBB. Also, the intestinal microbiome dysbiosis promotes the escape of pathogens into the circulatory system and CNS leading to the perpetual deterioration of neurons [115]. Though several animal model studies have proved it, more solid evidence is needed to affirm that modulation in the intestinal gut microflora is linked to psychological illnesses [116].

As discussed earlier, AD stems from atrophy of the cortex, neuronal loss, and plaque deposition caused due to amyloid peptide aggregation and tangled neurofibrils stormed by tau. Microbes are present all through the body with a variation in the type of metabolites produced by them [117]. In AD, aging is characterized not only by molecular and cellular modification but also by microbiome composition in the gut [118]. It is well studied that AD damages the BBB [14]. The combination of altered permeability at the intestinal tract triggers the pathogenic signalling between gut microbiota and the CNS owed to the 'gut-brain axis' phenomenon. The microbial dysbiosis alters gut permeability enabling a trigger to the systemic inflammatory status that could catalyze AD development [15]. It has been reflected in various studies that several pathogens like *Staphylococcus aureus, Mycobacterium tuberculosis, Salmonella enterica, Salmonella typhimurium, Bacillus subtilis, and Escherichia coli* have shown the ability to produce amyloid protein [17, 18]. *Porphyromona gingivalis* produces toxic proteases, gingipains, which were traced in the brain autopsy of the ailing from AD and have shown impairment of tau protein leading to neuronal dysfunction [7, 119].

However, it has also been observed that some microbes show a positive impact on reducing AD-associated symptoms. Several studies have proven the hypothesis that probiotics and prebiotics may have beneficial effects on the amelioration of AD-associated alterations [120, 121, 122]. Administration of probiotics, other nutritional interventions, and dietary supplementation are under study for their potential in therapeutic roles [123]. Bacteria like *Lactobacillus* and *Bacillus* species produce acetylcholine neurotransmitters that deplete in AD-affected patients, which may have a beneficial effect [115]. The microflora in the elderly population beyond 60 years of age is rather limited and resistant, more sensitive to changes in environmental factors [115]. A decrease in the presence of beneficial bacteria like *Prevotella, Bacteroides, Faecalibacterium prausnitzii, Lactobacilli,* and abundance of *Ruminococcus and Enterobacteriaceae* was observed [8]. The growth behaviour of various bacterial species in the AD is revealed in Table **1**.

Higher bacterial levels have been observed in AD afflicted brains as compared to the brains of healthy individuals [49]. An estimate of 90% Aβ plaques in AD patients' brains consisted of Herpes Simplex Virus-1 (HSV-1) DNA, which aggravated dementia by approximately 2.56 folds [124]. A significant drop in *Bifidobacterium and Lactobacillus* has shown the occurrence of depression, along with cognitive impairment and synapse associated disorders [11]. Altered gut microbes community leads to increased permeability of the gut leading to increased activities of unwanted harmful microbes. Reports indicate that the growth of bacteria, including Pseudomonas, Bacillus, Streptomyces, Staphylococcus, *etc.*, may participate in the amyloid plaque deposition [13] (Fig. 1).

Therapeutic Role of Probiotics in AD

Probiotics impart a healthy effect on the host, whereas prebiotics such as dietary fibers promotes the flourishing of the gut microflora [100]. It is also suggested that the consumption of biologically active peptides in the form of dairy supplements propel improved memory retention capacity and blocks microglial function [125, 126]. There have been preclinical studies that suggest that a high-fat diet may significantly contribute to dementia by changing the gut microflora [127, 128]. It is believed that diet can play a major role as both the cause and treatment therapy in AD [129].

There are a number of possible mechanisms of probiotics action to ameliorate AD. One of them is SCFA, mainly acetate, butyrate, lactate, and propionate produced by probiotics microbes through the fermentation of dietary fibers [101]. The SCFAs are known to regulate microglia maturation and functions. Further shreds of evidence indicate that probiotics act as strong anti-inflammatory and antioxidant agents to decrease neuroinflammation and oxidative stress, thus improving AD pathology [130].

Prebiotics For AD

Mediterranean diet prefers consumption of fruits, vegetables, cereals, legumes, olive oil, and adequate consumption of fish, cheese, and yogurt with a moderate amount of wine usually taken along with meals and rather low consumption of poultry and meat [131]. These components in the diet, inclusive of a healthy lifestyle, could reduce the susceptibility of AD and dementia [132, 133]. Studies have shown that dietary factors increase the chance of cardiovascular diseases, which is assumed to encourage the risk of dementia [134]. High consumption of fish pertains to a reduced risk of AD as the oils and fats from fish improve cerebrovascular activity. Thus, adapting to the Mediterranean diet could demote the decline of cognitive aspects, with an inclination towards the lesser risk of developing AD [135].

Many studied models and theories have indicated that higher consumption of prebiotics and probiotics along with other nutrients and lots of dietary fibre could lower the chances of developing AD [136]. The effects of the 'Mediterranean diet' on the host have been shown to propel a significant anti-inflammatory effect of gut bacteria [137]. However, the extent to which bacterial composition plays a role in psychological well-being affects these diets and improving AD symptoms remain unknown [138].

Gut Microbiome and Parkinson's Disease

The role of bidirectional communication between the gut-brain and gut microbiome in GI pathogenesis has received quite some attention in the case of PD [12, 139]. Many studies have testified that features a vital role in worsening the effects of PD [140, 141, 142]. The research revealed depletion of *prevotellaceae* while a hike in *Enterobacteriaceae* was observed in stool samples of PD patients [139], suggesting its possible connection with PD pathogenesis and disease progression. *Prevotellaceace* is commensals associated with mucin and SCFA producers *via* dietary fiber fermentation, and a decrease in their population would lead to increased gut permeability. This condition increases the probabilities of exposure to the bacterial endotoxins that inhibit the expression of α-synuclein in the gut whilst supporting its misfolding [143, 144]. It is also deduced that increase in *Enterobacteriaceae* elevates the Lipopolysaccharide (LPS) in serum from Gram-negative bacterial cell wall [143], which along with other neurotoxins, traverse the intestinal walls and enter the bloodstream. This transgression leads to the destruction of the intestinal epithelium barriers [145]. These elevated endotoxins and pathogens cause BBB disruption, which supports the deposition of α- synuclein [146, 147, 148]. A reduction in the number of bacteria that produce butyrates like *Roseburia, Coprococcus, Faecalibacterium,* and *Blautia* show anti-inflammatory properties, have been noted in stool samples of PD patients' [149]. The excessive bacterial growth which occurs in the small intestine is known as small intestine bacterial overgrowth (SIBO) and is speculated to be linked to PD [150]. It is proposed that byproducts of microbiota metabolism are SCFA which are capable of tweaking the ENS function, thus modulating gastrointestinal transit and motility [149]. However, the primary link between microbial dysbiosis and the progression of PD is yet to be learned more intensively. Indeed, many studies have inferred that there perhaps is a definite link between the development of PD and dysbiosis of the microbiome [151, 152, 153]. The growth behaviour of various bacterial species in the PD subjects is shown in Table **1**.

Probiotics and Prebiotics for PD

Lactobacilli, Enterococci, Bifidobacterium, and a mixture of different beneficial bacteria comprise the commonly used probiotics [154]. Data published to date is very poor to prove the direct beneficial effects of probiotics for PD. However, certain studies on animal models partially suggest that probiotics such as *Lactobacillus* could lessen LPS-induced neuroinflammation and further recover memory deficits [155] (Fig. **1**).

Table 1. **Behaviour of Gut Microbiota in Alzheimer's and Parkinson's disease.**

Gut Microbes	Alterations	References
Gut Microbiota in Alzheimer's Disease Condition		
Lactobacilli, Bacteroides, Prevotella,Clostridiaceae, Mogibacteriaceae, Bifidobacteriaceae, Faecalibacterium prausnitzii	Decrease in population	[7, 8, 14, 115]
Enterobacteria, Streptococci, Staphylococci, Chlamydia pneumonia. Enterobacteriaceae, Ruminococcus, Porphyromona gingivalis, Staphylococcus aureus, Mycobacterium tuberculosis, typhimurium, Bacillus subtilis, and yeast.	Increase in population	[7, 8, 11, 17, 18]
Gut Microbiota in Parkinson's Disease Condition		
B. coagulans, F. prausnitzii, Entercoccaceae, Enterobacteriaceae, Coprococcus, Propionobacteria	Decrease in population	[139, 151]
Lactobacillus, Bifidobacterium, S. boulardii, Ruminococcus, Blautia(butyrate produce), Faecalibacterium, Lachnospiraceae (SCFA producer)	Decrease in population	[9, 151, 152]
Clostridium coccoides, Bacteroides fragilis, Prevotellaceae, Erysipelotrichaceae, Enterococcaceae, Roseburia	Decrease in population	[9, 12, 139]
Escherichia, Shigella, Proteus, S. aureus, H. pylori, Prevotella, Streptococcus	Increase in population	[10, 12]

Mediterranean Diet

A Mediterranean diet comprises of daily consumption of whole grains, fruits, vegetables, nuts, and healthy fats with weekly consumption of poultry, fish, beans, and eggs. Moderate consumption of the dairy product and limited intake of red meat has been known to reduce the risk of PD as they are fiber-enriched diets and promote the growth of beneficial microbiome [114, 156, 157].

Ketogenic Diet and Fasting

Lifestyle and dietary interventions like imbibing a ketogenic diet and caloric restriction/fasting are currently used to treat neurological diseases. These effects are perhaps because ketosis increases neurotrophic factors like Brain-derived neurotrophic factor (BDNF), elevated levels of antioxidants, and reduces pro-inflammatory cytokine production. This is so widely sought after as the ketone bodies can cross the BBB and may bypass the type 1 complex defect of mitochondria in PD to rescue mitochondrial ATP function [157]. Additionally, ketone bodies, along with fasting and consuming a ketogenic diet, can impact PD positively by reducing insulin resistance and BDNF production [157]. Despite the promising possibilities reflected by this regime, well-designed trials are needed to be conducted to show if the ketogenic diet is beneficial in PD.

Polyphenols

Natural compounds like polyphenols and phenolics are organic compounds found in foods like tea, coffee, fruits, cereals, vegetables, and wine, *etc.* They modulate the composition of the gut microbial population by the inhibition of pathogenic bacteria and stimulation of beneficial bacteria. They possess a tendency to act as prebiotic compounds and enrich the beneficial bacteria [12]. Polyphenols have been known to show their neuroprotective potential to protect neurons against injury induced by neurotoxins by suppressing neuroinflammation and the potential to promote memory, learning, and cognitive function [158]. They have also been studied to show a significant drop in the levels of pathogens like *S. aureus, E. coli, H. Pylori, and Prevotella* through antibacterial properties [10, 12]. Hence, increasing the consumption of vegetables and fruits could facilitate the prevention of PD *via* positive effects on the intestinal microbiota composition [159].

CONCLUDING REMARKS

Research reviewed in the present chapter indicates the strong relation of gut microflora in normal health and disease pathogenesis. Several studies proved that there is a drastic change in the microbial community in the gut of AD as well as PD patients, and this could be a promising method for early diagnosis of these diseases. Amyloid protein, toxic proteases, and pro-inflammatory markers produced by pathogenic microbes lead to brain dysfunction accompanying by cognitive impairment and depression in AD. However, exposure to the bacterial endotoxins due to increased gut permeability and Gram-negative bacteria-derived LPS augments the likelihoods of α-synuclein expression and misfolding in the case of PD. Overall, gut dysbiosis causes neuroinflammation, oxidative stress and further increases the chance of disease generation due to environmental risk factors.

To ameliorate this condition, the use of probiotics can be a better treatment strategy. Probiotics could efficiently improve dementia and cognitive impairment in AD and PD. Maintenance of gut microbiota homeostasis using probiotics and prebiotics can improve the pathological factors involved in the progression of these NDDs as a novel therapeutic approach. The future research directions should, therefore, emphasize the exploration of gut microbiota for the purpose of disease diagnosis and further balancing proper gut microbiota utilizing probiotics and prebiotics as an empirical treatment approach.

CONSENT FOR PUBLICATION

Not applicable.

CONFLICT OF INTEREST

The authors declare no conflict of interest, financial or otherwise.

ACKNOWLEDGEMENTS

Declared none.

REFERENCES

[1]   a) Gao H-M, Hong J-S. Why neurodegenerative diseases are progressive: uncontrolled inflammation drives disease progression. Trends in Immunology 2008; 29(8): 357-365. b) Green R.C., Cupples L.A., Go R., Benke K.S., Edeki T., Griffith P.A., Williams M., Hipps Y., Graff-Radford N., Bachman D. Farrer L.A. Risk of dementia among white and African American relatives of patients with Alzheimer disease 2002.

[2]   Holtzman DM, Morris JC, Goate AM. Alzheimer's disease: the challenge of the second century. Sci Transl Med 2011; 3(77): 77sr1.
[http://dx.doi.org/10.1126/scitranslmed.3002369] [PMID: 21471435]

[3]   Bacskai BJ, Kajdasz ST, McLellan ME, *et al.* Non-Fc-mediated mechanisms are involved in clearance of amyloid-β *in vivo* by immunotherapy. J Neurosci 2002; 22(18): 7873-8.
[http://dx.doi.org/10.1523/JNEUROSCI.22-18-07873.2002] [PMID: 12223540]

[4]   Nizynski B, Dzwolak W, Nieznanski K. Amyloidogenesis of Tau protein. Protein Sci 2017; 26(11): 2126-50.
[http://dx.doi.org/10.1002/pro.3275] [PMID: 28833749]

[5]   World Health Organization; CC BY-NC-SA 3.0 IGO.. Risk Reduction of Cognitive Decline and Dementia 2019.

[6]   Politis M, Niccolini F. Serotonin in Parkinson's disease. Behav Brain Res 2015; 277(277): 136-45.
[http://dx.doi.org/10.1016/j.bbr.2014.07.037] [PMID: 25086269]

[7]   Bu XL, Yao XQ, Jiao SS, *et al.* A study on the association between infectious burden and Alzheimer's disease. Eur J Neurol 2015; 22(12): 1519-25.

[http://dx.doi.org/10.1111/ene.12477] [PMID: 24910016]

[8]   Cresci GA, Bawden E. Gut microbiome: what we do and don't know. Nutr Clin Pract 2015; 30(6): 734-46.
[http://dx.doi.org/10.1177/0884533615609899] [PMID: 26449893]

[9]   Hill D, Sugrue I, Arendt E, Hill C, Stanton C, Ross RP. Recent advances in microbial fermentation for dairy and health. F1000 Res 2017; 6: 751.
[http://dx.doi.org/10.12688/f1000research.10896.1] [PMID: 28649371]

[10]   Huang Y, Deng L, Zhong Y, Yi M. The Association between E326K of *GBA* and the Risk of Parkinson's Disease. Parkinsons Dis 2018; 2018: 1048084.
[http://dx.doi.org/10.1155/2018/1048084] [PMID: 29808112]

[11]   Strandwitz P. Neurotransmitter modulation by the gut microbiota. Brain Res 2018; 1693(Pt B): 128-33.
[http://dx.doi.org/10.1016/j.brainres.2018.03.015] [PMID: 29903615]

[12]   Unger MM, Spiegel J, Dillmann KU, *et al.* Short chain fatty acids and gut microbiota differ between patients with Parkinson's disease and age-matched controls. Parkinsonism Relat Disord 2016; 32: 66-72.
[http://dx.doi.org/10.1016/j.parkreldis.2016.08.019] [PMID: 27591074]

[13]   Zhang L, Wang Y, Xiayu X, *et al.* Altered gut microbiota in a mouse model of Alzheimer's disease. J Alzheimers Dis 2017; 60(4): 1241-57.
[http://dx.doi.org/10.3233/JAD-170020] [PMID: 29036812]

[14]   Calsolaro V, Edison P. Neuroinflammation in Alzheimer's disease: Current evidence and future directions. Alzheimers Dement 2016; 12(6): 719-32.
[http://dx.doi.org/10.1016/j.jalz.2016.02.010] [PMID: 27179961]

[15]   Spielman LJ, Gibson DL, Klegeris A. Unhealthy gut, unhealthy brain: The role of the intestinal microbiota in neurodegenerative diseases. Neurochem Int 2018; 120: 149-63.
[http://dx.doi.org/10.1016/j.neuint.2018.08.005] [PMID: 30114473]

[16]   Westfall S, Lomis N, Kahouli I, Dia SY, Singh SP, Prakash S. Microbiome, probiotics and neurodegenerative diseases: deciphering the gut brain axis. Cell Mol Life Sci 2017; 74(20): 3769-87.
[http://dx.doi.org/10.1007/s00018-017-2550-9] [PMID: 28643167]

[17]   Larsen N, Vogensen FK, van den Berg FW, *et al.* Gut microbiota in human adults with type 2 diabetes differs from non-diabetic adults. PLoS One 2010; 5(2): e9085.
[http://dx.doi.org/10.1371/journal.pone.0009085] [PMID: 20140211]

[18]   Oli MW, Otoo HN, Crowley PJ, *et al.* Functional amyloid formation by Streptococcus mutans. Microbiology (Reading) 2012; 158(Pt 12): 2903-16.
[http://dx.doi.org/10.1099/mic.0.060855-0] [PMID: 23082034]

[19]   Nair AT, Ramachandran V, Joghee NM, Antony S, Ramalingam G. Gut Microbiota Dysfunction as Reliable Non-invasive Early Diagnostic Biomarkers in the Pathophysiology of Parkinson's Disease: A Critical Review. J Neurogastroenterol Motil 2018; 24(1): 30-42.
[http://dx.doi.org/10.5056/jnm17105] [PMID: 29291606]

[20]   Li Z, Zhu H, Zhang L, Qin C. The intestinal microbiome and Alzheimer's disease: A review. Animal models and experimental medicine 2018; 1(3): 180-8.
[http://dx.doi.org/10.1002/ame2.12033]

[21]   Cummings JL, Cole G. Alzheimer disease. JAMA 2002; 287(18): 2335-8.
[http://dx.doi.org/10.1001/jama.287.18.2335] [PMID: 11988038]

[22]   Bramblett GT, Goedert M, Jakes R, Merrick SE, Trojanowski JQ, Lee VM. Abnormal tau phosphorylation at Ser396 in Alzheimer's disease recapitulates development and contributes to reduced microtubule binding. Neuron 1993; 10(6): 1089-99.
[http://dx.doi.org/10.1016/0896-6273(93)90057-X] [PMID: 8318230]

[23] Holmes C, Boche D, Wilkinson D, *et al.* Long-term effects of Abeta42 immunisation in Alzheimer's disease: follow-up of a randomised, placebo-controlled phase I trial. Lancet 2008; 372(9634): 216-23. [http://dx.doi.org/10.1016/S0140-6736(08)61075-2] [PMID: 18640458]

[24] Chen JJ, Zhao B, Zhao J, Li S. Potential roles of exosomal microRNAs as diagnostic biomarkers and therapeutic application in Alzheimer's disease. Neural Plast 2017; 2017: 7027380. [http://dx.doi.org/10.1155/2017/7027380] [PMID: 28770113]

[25] Stockley JH, O'Neill C. Understanding BACE1: essential protease for amyloid-β production in Alzheimer's disease. Cell Mol Life Sci 2008; 65(20): 3265-89. [http://dx.doi.org/10.1007/s00018-008-8271-3] [PMID: 18695942]

[26] Govindpani K, McNamara LG, Smith NR, *et al.* Vascular Dysfunction in Alzheimer's Disease: A Prelude to the Pathological Process or a Consequence of It? J Clin Med 2019; 8(5): 651. [http://dx.doi.org/10.3390/jcm8050651] [PMID: 31083442]

[27] Hampel H, Mesulam MM, Cuello AC, *et al.* The cholinergic system in the pathophysiology and treatment of Alzheimer's disease. Brain 2018; 141(7): 1917-33. [http://dx.doi.org/10.1093/brain/awy132] [PMID: 29850777]

[28] Brookmeyer R, Johnson E, Ziegler-Graham K, Arrighi HM. Forecasting the global burden of Alzheimer's disease. Alzheimers Dement 2007; 3(3): 186-91. [http://dx.doi.org/10.1016/j.jalz.2007.04.381] [PMID: 19595937]

[29] Ogeng'o JA, Cohen DL, Sayi JG, *et al.* Cerebral amyloid β protein deposits and other Alzheimer lesions in non-demented elderly east Africans. Brain Pathol 1996; 6(2): 101-7. [http://dx.doi.org/10.1111/j.1750-3639.1996.tb00790.x] [PMID: 8737923]

[30] Schneider LS, Sano M. Current Alzheimer's disease clinical trials: methods and placebo outcomes. Alzheimers Dement 2009; 5(5): 388-97. [http://dx.doi.org/10.1016/j.jalz.2009.07.038] [PMID: 19751918]

[31] Jellinger KA, Attems J. Neurofibrillary tangle-predominant dementia: comparison with classical Alzheimer disease. Acta Neuropathol 2007; 113(2): 107-17. [http://dx.doi.org/10.1007/s00401-006-0156-7] [PMID: 17089134]

[32] Lopez OL, Becker JT, Sweet RA, *et al.* Psychiatric symptoms vary with the severity of dementia in probable Alzheimer's disease. J Neuropsychiatry Clin Neurosci 2003; 15(3): 346-53. [http://dx.doi.org/10.1176/jnp.15.3.346] [PMID: 12928511]

[33] Petersen RC, Smith GE, Waring SC, Ivnik RJ, Tangalos EG, Kokmen E. Mild cognitive impairment: clinical characterization and outcome. Arch Neurol 1999; 56(3): 303-8. [http://dx.doi.org/10.1001/archneur.56.3.303] [PMID: 10190820]

[34] Daviglus ML, Bell CC, Berrettini W, *et al.* National Institutes of Health State-of-the-Science Conference statement: preventing alzheimer disease and cognitive decline. Ann Intern Med 2010; 153(3): 176-81. [http://dx.doi.org/10.7326/0003-4819-153-3-201008030-00260] [PMID: 20547888]

[35] Fratiglioni L, Ahlbom A, Viitanen M, Winblad B. Risk factors for late-onset Alzheimer's disease: a population-based, case-control study. Ann Neurol 1993; 33(3): 258-66. [http://dx.doi.org/10.1002/ana.410330306] [PMID: 8498809]

[36] Mayeux R, Sano M, Chen J, Tatemichi T, Stern Y. Risk of dementia in first-degree relatives of patients with Alzheimer's disease and related disorders. Arch Neurol 1991; 48(3): 269-73. [http://dx.doi.org/10.1001/archneur.1991.00530150037014] [PMID: 2001183]

[37] McKhann GM, Knopman DS, Chertkow H, *et al.* The diagnosis of dementia due to Alzheimer's disease: recommendations from the National Institute on Aging-Alzheimer's Association workgroups on diagnostic guidelines for Alzheimer's disease. Alzheimers Dement 2011; 7(3): 263-9. [http://dx.doi.org/10.1016/j.jalz.2011.03.005] [PMID: 21514250]

[38]   Lautenschlager NT, Cupples LA, Rao VS, *et al.* Risk of dementia among relatives of Alzheimer's disease patients in the MIRAGE study: What is in store for the oldest old? Neurology 1996; 46(3): 641-50.
[http://dx.doi.org/10.1212/WNL.46.3.641] [PMID: 8618660]

[39]   James BD, Boyle PA, Buchman AS, Barnes LL, Bennett DA. Life space and risk of Alzheimer disease, mild cognitive impairment, and cognitive decline in old age. Am J Geriatr Psychiatry 2011; 19(11): 961-9.
[http://dx.doi.org/10.1097/JGP.0b013e318211c219] [PMID: 21430509]

[40]   Pákáski M, Kálmán J. Interactions between the amyloid and cholinergic mechanisms in Alzheimer's disease. Neurochem Int 2008; 53(5): 103-11.
[http://dx.doi.org/10.1016/j.neuint.2008.06.005] [PMID: 18602955]

[41]   Perry RJ, Hodges JR. Attention and executive deficits in Alzheimer's disease. A critical review. Brain 1999; 122(Pt 3): 383-404.
[http://dx.doi.org/10.1093/brain/122.3.383] [PMID: 10094249]

[42]   Sarter M, Hasselmo ME, Bruno JP, Givens B. Unraveling the attentional functions of cortical cholinergic inputs: interactions between signal-driven and cognitive modulation of signal detection. Brain Res Brain Res Rev 2005; 48(1): 98-111.
[http://dx.doi.org/10.1016/j.brainresrev.2004.08.006] [PMID: 15708630]

[43]   Dineley KT. Beta-amyloid peptide--nicotinic acetylcholine receptor interaction: the two faces of health and disease. Front Biosci 2007; 12: 5030-8.
[http://dx.doi.org/10.2741/2445] [PMID: 17569627]

[44]   Banerjee C, Nyengaard JR, Wevers A, *et al.* Cellular expression of alpha7 nicotinic acetylcholine receptor protein in the temporal cortex in Alzheimer's and Parkinson's disease--a stereological approach. Neurobiol Dis 2000; 7(6 Pt B): 666-72.
[http://dx.doi.org/10.1006/nbdi.2000.0317] [PMID: 11114264]

[45]   Liu Q, Wu J. Neuronal nicotinic acetylcholine receptors serve as sensitive targets that mediate beta-amyloid neurotoxicity. Acta Pharmacol Sin 2006; 27(10): 1277-86.
[http://dx.doi.org/10.1111/j.1745-7254.2006.00430.x] [PMID: 17007734]

[46]   Barrantes FJ, Borroni V, Vallés S. Neuronal nicotinic acetylcholine receptor-cholesterol crosstalk in Alzheimer's disease. FEBS Lett 2010; 584(9): 1856-63.
[http://dx.doi.org/10.1016/j.febslet.2009.11.036] [PMID: 19914249]

[47]   Iyo M, Namba H, Fukushi K, *et al.* Measurement of acetylcholinesterase by positron emission tomography in the brains of healthy controls and patients with Alzheimer's disease. Lancet 1997; 349(9068): 1805-9.
[http://dx.doi.org/10.1016/S0140-6736(96)09124-6] [PMID: 9269216]

[48]   Jagtap U, Lekhak M, Fulzele D, Yadav S, Bapat V. Analysis of selected Crinum species for galanthamine alkaloid: an anti-Alzheimer drug. Curr Sci 2014; 107(12): 2008-10.

[49]   Emery DC, Shoemark DK, Batstone TE, *et al.* 16S rRNA next generation sequencing analysis shows bacteria in Alzheimer's post-mortem brain. Frontiers in aging neuroscience, 9, p.195. Esler, W.P. and Wolfe, M.S., 2001. A portrait of Alzheimer secretases--new features and familiar faces. Science 2017; 293(5534): 1449-54.

[50]   Cai H, Wang Y, McCarthy D, *et al.* BACE1 is the major β-secretase for generation of Abeta peptides by neurons. Nat Neurosci 2001; 4(3): 233-4.
[http://dx.doi.org/10.1038/85064] [PMID: 11224536]

[51]   Selkoe DJ. Alzheimer's disease: genes, proteins, and therapy. Physiol Rev 2001; 81(2): 741-66.
[http://dx.doi.org/10.1152/physrev.2001.81.2.741] [PMID: 11274343]

[52]   van Dyck CH. Anti-amyloid-β monoclonal antibodies for Alzheimer's disease: pitfalls and promise. Biol Psychiatry 2018; 83(4): 311-9.

[http://dx.doi.org/10.1016/j.biopsych.2017.08.010] [PMID: 28967385]

[53] Dodart JC, Bales KR, Gannon KS, *et al.* Immunization reverses memory deficits without reducing brain Abeta burden in Alzheimer's disease model. Nat Neurosci 2002; 5(5): 452-7.
[http://dx.doi.org/10.1038/nn842] [PMID: 11941374]

[54] Permanne B, Adessi C, Saborio GP, *et al.* Reduction of amyloid load and cerebral damage in a transgenic mouse model of Alzheimer's disease by treatment with a β-sheet breaker peptide. FASEB J 2002; 16(8): 860-2.
[http://dx.doi.org/10.1096/fj.01-0841fje] [PMID: 11967228]

[55] Stewart WF, Kawas C, Corrada M, Metter EJ. Risk of Alzheimer's disease and duration of NSAID use. Neurology 1997; 48(3): 626-32.
[http://dx.doi.org/10.1212/WNL.48.3.626] [PMID: 9065537]

[56] Ho L, Purohit D, Haroutunian V, *et al.* Neuronal cyclooxygenase 2 expression in the hippocampal formation as a function of the clinical progression of Alzheimer disease. Arch Neurol 2001; 58(3): 487-92.
[http://dx.doi.org/10.1001/archneur.58.3.487] [PMID: 11255454]

[57] Kojro E, Fahrenholz F. The non-amyloidogenic pathway: structure and function of α-secretases.Alzheimer's disease. Boston, MA: Springer 2005; pp. 105-27.

[58] Fassbender K, Simons M, Bergmann C, *et al.* Simvastatin strongly reduces levels of Alzheimer's disease β -amyloid peptides Abeta 42 and Abeta 40 *in vitro* and *in vivo*. Proc Natl Acad Sci USA 2001; 98(10): 5856-61.
[http://dx.doi.org/10.1073/pnas.081620098] [PMID: 11296263]

[59] Wolozin B, Kellman W, Ruosseau P, Celesia GG, Siegel G. Decreased prevalence of Alzheimer disease associated with 3-hydroxy-3-methyglutaryl coenzyme A reductase inhibitors. Arch Neurol 2000; 57(10): 1439-43.
[http://dx.doi.org/10.1001/archneur.57.10.1439] [PMID: 11030795]

[60] Galimberti D, Schoonenboom N, Scarpini E, Scheltens P. Dutch-Italian Alzheimer Research Group. Chemokines in serum and cerebrospinal fluid of Alzheimer's disease patients. Ann Neurol 2003; 53(4): 547-8.
[http://dx.doi.org/10.1002/ana.10531] [PMID: 12666129]

[61] Roberts GW, Gentleman SM, Lynch A, Murray L, Landon M, Graham DI. Beta amyloid protein deposition in the brain after severe head injury: implications for the pathogenesis of Alzheimer's disease. J Neurol Neurosurg Psychiatry 1994; 57(4): 419-25.
[http://dx.doi.org/10.1136/jnnp.57.4.419] [PMID: 8163989]

[62] Carpenter BD, Strauss ME, Kennedy JS. Personal history of depression and its appearance in Alzheimer's disease. Int J Geriatr Psychiatry 1995; 10(8): 669-78.
[http://dx.doi.org/10.1002/gps.930100807]

[63] Funke C, Schneider SA, Berg D, Kell DB. Genetics and iron in the systems biology of Parkinson's disease and some related disorders. Neurochem Int 2013; 62(5): 637-52.
[http://dx.doi.org/10.1016/j.neuint.2012.11.015] [PMID: 23220386]

[64] Greenamyre JT, Hastings TG. Biomedicine. Parkinson's--divergent causes, convergent mechanisms. Science 2004; 304(5674): 1120-2.
[http://dx.doi.org/10.1126/science.1098966] [PMID: 15155938]

[65] de Rijk MC, Tzourio C, Breteler MM, *et al.* Prevalence of parkinsonism and Parkinson's disease in Europe: the EUROPARKINSON Collaborative Study. European Community Concerted Action on the Epidemiology of Parkinson's disease. J Neurol Neurosurg Psychiatry 1997; 62(1): 10-5.
[http://dx.doi.org/10.1136/jnnp.62.1.10] [PMID: 9010393]

[66] Litvan I, Bhatia KP, Burn DJ, *et al.* Movement Disorders Society Scientific Issues Committee. Movement Disorders Society Scientific Issues Committee report: SIC Task Force appraisal of clinical

diagnostic criteria for Parkinsonian disorders. Mov Disord 2003; 18(5): 467-86.
[http://dx.doi.org/10.1002/mds.10459] [PMID: 12722160]

[67]   de Lau LM, Breteler MM. Epidemiology of Parkinson's disease. Lancet Neurol 2006; 5(6): 525-35.
[http://dx.doi.org/10.1016/S1474-4422(06)70471-9] [PMID: 16713924]

[68]   Mizuno Y, Hattori N, Mori H, Suzuki T, Tanaka K. Parkin and Parkinson's disease. Curr Opin Neurol 2001; 14(4): 477-82.
[http://dx.doi.org/10.1097/00019052-200108000-00008] [PMID: 11470964]

[69]   Biswas A, Gupta A, Naiya T, *et al.* Molecular pathogenesis of Parkinson's disease: identification of mutations in the Parkin gene in Indian patients. Parkinsonism Relat Disord 2006; 12(7): 420-6.
[http://dx.doi.org/10.1016/j.parkreldis.2006.04.005] [PMID: 16793319]

[70]   Dorsey ER, Elbaz A, Nichols E, *et al.* GBD 2016 Parkinson's Disease Collaborators. Global, regional, and national burden of Parkinson's disease, 1990-2016: a systematic analysis for the Global Burden of Disease Study 2016. Lancet Neurol 2018; 17(11): 939-53.
[http://dx.doi.org/10.1016/S1474-4422(18)30295-3] [PMID: 30287051]

[71]   Bellou V, Belbasis L, Tzoulaki I, Evangelou E, Ioannidis JP. Environmental risk factors and Parkinson's disease: An umbrella review of meta-analyses. Parkinsonism Relat Disord 2016; 23: 1-9.
[http://dx.doi.org/10.1016/j.parkreldis.2015.12.008] [PMID: 26739246]

[72]   Harris H, Rubinsztein DC. Control of autophagy as a therapy for neurodegenerative disease. Nat Rev Neurol 2011; 8(2): 108-17.
[http://dx.doi.org/10.1038/nrneurol.2011.200] [PMID: 22187000]

[73]   Dal Forno G, Rasmusson DX, Brandt J, *et al.* Apolipoprotein E genotype and rate of decline in probable Alzheimer's disease. Arch Neurol 1996; 53(4): 345-50.
[http://dx.doi.org/10.1001/archneur.1996.00550040085017] [PMID: 8929157]

[74]   Lewis GN, Byblow WD, Walt SE. Stride length regulation in Parkinson's disease: the use of extrinsic, visual cues. Brain 2000; 123(Pt 10): 2077-90.
[http://dx.doi.org/10.1093/brain/123.10.2077] [PMID: 11004125]

[75]   Spillantini MG, Crowther RA, Jakes R, Hasegawa M, Goedert M. α-Synuclein in filamentous inclusions of Lewy bodies from Parkinson's disease and dementia with lewy bodies. Proc Natl Acad Sci USA 1998; 95(11): 6469-73.
[http://dx.doi.org/10.1073/pnas.95.11.6469] [PMID: 9600990]

[76]   Charvin D, Medori R, Hauser RA, Rascol O. Therapeutic strategies for Parkinson disease: beyond dopaminergic drugs. Nat Rev Drug Discov 2018; 17(11): 804-22.
[http://dx.doi.org/10.1038/nrd.2018.136] [PMID: 30262889]

[77]   Lawrence AD, Evans AH, Lees AJ. Compulsive use of dopamine replacement therapy in Parkinson's disease: reward systems gone awry? Lancet Neurol 2003; 2(10): 595-604.
[http://dx.doi.org/10.1016/S1474-4422(03)00529-5] [PMID: 14505581]

[78]   Cohen D, Nabirochkin S, Chumakov I, Hajj R. Therapeutic approaches for treating Parkinson's disease. U.S. Patent 10,010,515, 2018.

[79]   Kim HJ, Jeon BS, Jenner P. Hallmarks of treatment aspects: Parkinson's disease throughout centuries including l-Dopa. Int Rev Neurobiol 2017; 132: 295-343.
[http://dx.doi.org/10.1016/bs.irn.2017.01.006] [PMID: 28554412]

[80]   Korczyn AD, Nussbaum M. Emerging therapies in the pharmacological treatment of Parkinson's disease. Drugs 2002; 62(5): 775-86.
[http://dx.doi.org/10.2165/00003495-200262050-00005] [PMID: 11929331]

[81]   Huot P, Johnston TH, Koprich JB, Fox SH, Brotchie JM. The pharmacology of L-DOPA-induced dyskinesia in Parkinson's disease. Pharmacol Rev 2013; 65(1): 171-222.
[http://dx.doi.org/10.1124/pr.111.005678] [PMID: 23319549]

[82]   Kostrzewa RM, Kostrzewa JP, Brus R. Neuroprotective and neurotoxic roles of levodopa (L-DOPA) in neurodegenerative disorders relating to Parkinson's disease. Amino Acids 2002; 23(1-3): 57-63.
[http://dx.doi.org/10.1007/s00726-001-0110-x] [PMID: 12373519]

[83]   Patil R, Aware C, Gaikwad S, *et al.* RP-HPLC Analysis of Anti-Parkinson's Drug L-DOPA Content in Mucuna Species from Indian Subcontinent. Proc Natl Acad Sci, India, Sect B Biol Sci 2019; 89: 1413-20.
[http://dx.doi.org/10.1007/s40011-018-01071-9]

[84]   Patil R, Gholave A, Yadav S, Bapat V, Jadhav J. Mucuna sanjappae Aitawade et Yadav: a new species of Mucuna with promising yield of anti- Parkinson"s drug L-DOPA. Genet Resour Crop Evol 2015; 62(1): 155-62.
[http://dx.doi.org/10.1007/s10722-014-0164-8]

[85]   Patil RR, Pawar KD, Rane MR, Yadav SR, Bapat VA, Jadhav JP. Assessment of genetic diversity in Mucuna species of India using randomly amplified polymorphic DNA and inter simple sequence repeat markers. Physiol Mol Biol Plants 2016; 22(2): 207-17.
[http://dx.doi.org/10.1007/s12298-016-0361-3] [PMID: 27436912]

[86]   Rai SN, Birla H, Singh SS, *et al. Mucuna pruriens* Protects against MPTP Intoxicated Neuroinflammation in Parkinson's Disease through NF-κB/pAKT Signaling Pathways. Front Aging Neurosci 2017; 9: 421.
[http://dx.doi.org/10.3389/fnagi.2017.00421] [PMID: 29311905]

[87]   Yadav SK, Prakash J, Chouhan S, Singh SP. Mucuna pruriens seed extract reduces oxidative stress in nigrostriatal tissue and improves neurobehavioral activity in paraquat-induced Parkinsonian mouse model. Neurochem Int 2013; 62(8): 1039-47.
[http://dx.doi.org/10.1016/j.neuint.2013.03.015] [PMID: 23562769]

[88]   Haq IU, Ali S. Mutation of Aspergillus oryzae for improved production of 3, 4-dihydroxy Phenyl--Alanine (L-Dopa) from L-Tyrosine. Braz J Microbiol 2006; 37(1): 78-86.
[http://dx.doi.org/10.1590/S1517-83822006000100015]

[89]   Mena MA, Casarejos MJ, Carazo A, Paíno CL, García de Yébenes J. Glia protect fetal midbrain dopamine neurons in culture from L-DOPA toxicity through multiple mechanisms. J Neural Transm (Vienna) 1997; 104(4-5): 317-28.
[http://dx.doi.org/10.1007/BF01277654] [PMID: 9295168]

[90]   Hirsch EC, Hunot S, Damier P, Faucheux B. Glial cells and inflammation in Parkinson's disease: a role in neurodegeneration? Annals of neurology 1998; 44(S1 1): S115-20.
[http://dx.doi.org/10.1002/ana.410440717]

[91]   Kringelbach ML, Jenkinson N, Owen SL, Aziz TZ. Translational principles of deep brain stimulation. Nat Rev Neurosci 2007; 8(8): 623-35.
[http://dx.doi.org/10.1038/nrn2196] [PMID: 17637800]

[92]   Quigley L, O'Sullivan O, Stanton C, *et al.* The complex microbiota of raw milk. FEMS Microbiol Rev 2013; 37(5): 664-98.
[http://dx.doi.org/10.1111/1574-6976.12030] [PMID: 23808865]

[93]   Fouhy F, Guinane CM, Hussey S, *et al.* High-throughput sequencing reveals the incomplete, short-term recovery of infant gut microbiota following parenteral antibiotic treatment with ampicillin and gentamicin. Antimicrob Agents Chemother 2012; 56(11): 5811-20.
[http://dx.doi.org/10.1128/AAC.00789-12] [PMID: 22948872]

[94]   Marques TM, Wall R, Ross RP, Fitzgerald GF, Ryan CA, Stanton C. Programming infant gut microbiota: influence of dietary and environmental factors. Curr Opin Biotechnol 2010; 21(2): 149-56.
[http://dx.doi.org/10.1016/j.copbio.2010.03.020] [PMID: 20434324]

[95]   Joshi D, Roy S, Banerjee S. Prebiotics: a functional food in health and disease. Natural products and drug discovery 2018; 507-23.

[96]   Guinane CM, Cotter PD. Role of the gut microbiota in health and chronic gastrointestinal disease: understanding a hidden metabolic organ. Therap Adv Gastroenterol 2013; 6(4): 295-308. [http://dx.doi.org/10.1177/1756283X13482996] [PMID: 23814609]

[97]   Frank DN, St Amand AL, Feldman RA, Boedeker EC, Harpaz N, Pace NR. Molecular-phylogenetic characterization of microbial community imbalances in human inflammatory bowel diseases. Proc Natl Acad Sci USA 2007; 104(34): 13780-5. [http://dx.doi.org/10.1073/pnas.0706625104] [PMID: 17699621]

[98]   Gupta V, Garg R. Probiotics. Indian J Med Microbiol 2009; 27(3): 202-9. [http://dx.doi.org/10.4103/0255-0857.53201] [PMID: 19584499]

[99]   Boaventura C, Azevedo R, Uetanabaro A, Nicoli J, Braga LG. The benefits of probiotics in human and animal nutrition. New Advances in the Basic and Clinical Gastroenterology 2012; 75. [http://dx.doi.org/10.5772/34027]

[100]   Camfield P, Camfield C. Transition to adult care for children with chronic neurological disorders. Ann Neurol 2011; 69(3): 437-44. [http://dx.doi.org/10.1002/ana.22393] [PMID: 21391239]

[101]   Koh A, De Vadder F, Kovatcheva-Datchary P, Bäckhed F. From dietary fiber to host physiology: short-chain fatty acids as key bacterial metabolites. Cell 2016; 165(6): 1332-45. [http://dx.doi.org/10.1016/j.cell.2016.05.041] [PMID: 27259147]

[102]   Belmaker RH, Agam G. Major depressive disorder. N Engl J Med 2008; 358(1): 55-68. [http://dx.doi.org/10.1056/NEJMra073096] [PMID: 18172175]

[103]   Collins SM, Bercik P. The relationship between intestinal microbiota and the central nervous system in normal gastrointestinal function and disease. Gastroenterology 2009; 136(6): 2003-14. [http://dx.doi.org/10.1053/j.gastro.2009.01.075] [PMID: 19457424]

[104]   Forsythe P, Bienenstock J, Kunze WA. Vagal pathways for microbiome-brain-gut axis communication. Microbial endocrinology: the microbiota-gut-brain axis in health and disease 2014; 115-33. [http://dx.doi.org/10.1007/978-1-4939-0897-4_5]

[105]   Falony G, Joossens M, Vieira-Silva S, *et al.* Population-level analysis of gut microbiome variation. Science 2016; 352(6285): 560-4. [http://dx.doi.org/10.1126/science.aad3503] [PMID: 27126039]

[106]   Ankolekar C, Johnson D, Pinto MdaS, Johnson K, Labbe R, Shetty K. Inhibitory potential of tea polyphenolics and influence of extraction time against Helicobacter pylori and lack of inhibition of beneficial lactic acid bacteria. J Med Food 2011; 14(11): 1321-9. [http://dx.doi.org/10.1089/jmf.2010.0237] [PMID: 21663484]

[107]   Jenkins TP, Peachey LE, Ajami NJ, *et al.* Schistosoma mansoni infection is associated with quantitative and qualitative modifications of the mammalian intestinal microbiota. Sci Rep 2018; 8(1): 12072. [http://dx.doi.org/10.1038/s41598-018-30412-x] [PMID: 30104612]

[108]   Pluta R, Ułamek-Kozioł M, Januszewski S, Czuczwar SJ. Gut microbiota and pro/prebiotics in Alzheimer's disease. Aging (Albany NY) 2020; 12(6): 5539-50. [http://dx.doi.org/10.18632/aging.102930] [PMID: 32191919]

[109]   Bäuerl C, Collado MC, Diaz Cuevas A, Viña J, Pérez Martínez G. Shifts in gut microbiota composition in an APP/PSS1 transgenic mouse model of Alzheimer's disease during lifespan. Lett Appl Microbiol 2018; 66(6): 464-71. [http://dx.doi.org/10.1111/lam.12882] [PMID: 29575030]

[110]   Girolamo F, Coppola C, Ribatti D. Immunoregulatory effect of mast cells influenced by microbes in neurodegenerative diseases. Brain Behav Immun 2017; 65: 68-89. [http://dx.doi.org/10.1016/j.bbi.2017.06.017] [PMID: 28676349]

[111] Pryde SE, Duncan SH, Hold GL, Stewart CS, Flint HJ. The microbiology of butyrate formation in the human colon. FEMS Microbiol Lett 2002; 217(2): 133-9.
[http://dx.doi.org/10.1111/j.1574-6968.2002.tb11467.x] [PMID: 12480096]

[112] Quigley EMM. Microbiota-brain-gut axis and neurodegenerative diseases. Curr Neurol Neurosci Rep 2017; 17(12): 94.
[http://dx.doi.org/10.1007/s11910-017-0802-6] [PMID: 29039142]

[113] Vital M, Howe AC, Tiedje JM. Revealing the bacterial butyrate synthesis pathways by analyzing (meta)genomic data. MBio 2014; 5(2): e00889-14.
[http://dx.doi.org/10.1128/mBio.00889-14] [PMID: 24757212]

[114] Makki K, Deehan EC, Walter J, Bäckhed F. The impact of dietary fiber on gut microbiota in host health and disease. Cell Host Microbe 2018; 23(6): 705-15.
[http://dx.doi.org/10.1016/j.chom.2018.05.012] [PMID: 29902436]

[115] Zuo T, Ng SC. The gut microbiota in the pathogenesis and therapeutics of inflammatory bowel disease. Front Microbiol 2018; 9: 2247.
[http://dx.doi.org/10.3389/fmicb.2018.02247] [PMID: 30319571]

[116] Hulme H, Meikle LM, Strittmatter N, et al. Microbiome-derived carnitine mimics as previously unknown mediators of gut-brain axis communication. Sci Adv 2020; 6(11): eaax6328.
[http://dx.doi.org/10.1126/sciadv.aax6328] [PMID: 32195337]

[117] Sender R, Fuchs S, Milo R. Are we really vastly outnumbered? Revisiting the ratio of bacterial to host cells in humans. Cell 2016; 164(3): 337-40.
[http://dx.doi.org/10.1016/j.cell.2016.01.013] [PMID: 26824647]

[118] Sarkar A, Harty S, Lehto SM, et al. The microbiome in psychology and cognitive neuroscience. Trends Cogn Sci 2018; 22(7): 611-36.
[http://dx.doi.org/10.1016/j.tics.2018.04.006] [PMID: 29907531]

[119] Singhrao SK, Harding A, Poole S, Kesavalu L, Crean S. Porphyromonas gingivalis periodontal infection and its putative links with Alzheimer's disease. Mediators Inflamm 2015; 2015: 137357.
[http://dx.doi.org/10.1155/2015/137357] [PMID: 26063967]

[120] Athari Nik Azm S, Djazayeri A, Safa M, et al. Lactobacilli and bifidobacteria ameliorate memory and learning deficits and oxidative stress in β-amyloid (1-42) injected rats. Appl Physiol Nutr Metab 2018; 43(7): 718-26.
[http://dx.doi.org/10.1139/apnm-2017-0648] [PMID: 29462572]

[121] Hosseinifard ES, Saghafi-Asl M, Bavafa-Valenlia K, Morshedi M. Correction to: The novel insight into anti-inflammatory and anxiolytic effects of psychobiotics in diabetic rats: possible link between gut microbiota and brain regions. Eur J Nutr 2019; 58(8): 3377-7.
[http://dx.doi.org/10.1007/s00394-019-02079-1] [PMID: 31492974]

[122] Sanborn V, Azcarate-Peril MA, Updegraff J, Manderino LM, Gunstad J. A randomized clinical trial examining the impact of LGG probiotic supplementation on psychological status in middle-aged and older adults. Contemp Clin Trials Commun 2018; 12: 192-7.
[http://dx.doi.org/10.1016/j.conctc.2018.11.006] [PMID: 30511028]

[123] Fernandez-Real JM, Serino M, Blasco G, et al. Gut microbiota interacts with brain microstructure and function. J Clin Endocrinol Metab 2015; 100(12): 4505-13.
[http://dx.doi.org/10.1210/jc.2015-3076] [PMID: 26445114]

[124] Tzeng NS, Chung CH, Lin FH, et al. Anti-herpetic medications and reduced risk of dementia in patients with herpes simplex virus infections—a nationwide, population-based cohort study in Taiwan. Neurotherapeutics 2018; 15(2): 417-29.
[http://dx.doi.org/10.1007/s13311-018-0611-x] [PMID: 29488144]

[125] Ano Y, Ayabe T, Kutsukake T, et al. Novel lactopeptides in fermented dairy products improve memory function and cognitive decline. Neurobiol Aging 2018; 72: 23-31.

[http://dx.doi.org/10.1016/j.neurobiolaging.2018.07.016] [PMID: 30176402]

[126]   Ano Y, Ayabe T, Ohya R, Kondo K, Kitaoka S, Furuyashiki T. Tryptophan-Tyrosine Dipeptide, the Core Sequence of β-Lactolin, Improves Memory by Modulating the Dopamine System. Nutrients 2019; 11(2): 348.
[http://dx.doi.org/10.3390/nu11020348] [PMID: 30736353]

[127]   Lee ES, Song EJ, Nam YD. Dysbiosis of gut microbiome and its impact on epigenetic regulation. J Clin Epigenet 2017; 1: 14.

[128]   Sanguinetti E, Collado MC, Marrachelli VG, *et al*. Microbiome-metabolome signatures in mice genetically prone to develop dementia, fed a normal or fatty diet. Sci Rep 2018; 8(1): 4907.
[http://dx.doi.org/10.1038/s41598-018-23261-1] [PMID: 29559675]

[129]   Solfrizzi V, Custodero C, Lozupone M, *et al*. Relationships of dietary patterns, foods, and micro-and macronutrients with Alzheimer's disease and late-life cognitive disorders: a systematic review. J Alzheimers Dis 2017; 59(3): 815-49.
[http://dx.doi.org/10.3233/JAD-170248] [PMID: 28697569]

[130]   Wong CH, Wu CY. Methods for modifying human antibodies by glycan engineering U.S. Patent 10,087,236, 2018.

[131]   Trichopoulou A, Costacou T, Bamia C, Trichopoulos D. Adherence to a Mediterranean diet and survival in a Greek population. N Engl J Med 2003; 348(26): 2599-608.
[http://dx.doi.org/10.1056/NEJMoa025039] [PMID: 12826634]

[132]   Frisardi V, Panza F, Seripa D, *et al*. Nutraceutical properties of Mediterranean diet and cognitive decline: possible underlying mechanisms. J Alzheimers Dis 2010; 22(3): 715-40.
[http://dx.doi.org/10.3233/JAD-2010-100942] [PMID: 20858954]

[133]   Scarmeas N, Stern Y, Tang M-X, Mayeux R, Luchsinger JA. Mediterranean diet and risk for Alzheimer's disease. Ann Neurol 2006; 59(6): 912-21.
[http://dx.doi.org/10.1002/ana.20854] [PMID: 16622828]

[134]   Kruman II, Culmsee C, Chan SL, *et al*. Homocysteine elicits a DNA damage response in neurons that promotes apoptosis and hypersensitivity to excitotoxicity. J Neurosci 2000; 20(18): 6920-6.
[http://dx.doi.org/10.1523/JNEUROSCI.20-18-06920.2000] [PMID: 10995836]

[135]   Sofi F, Cesari F, Abbate R, Gensini GF, Casini A. Adherence to Mediterranean diet and health status: metaanalysis. BMJ337 2008; a1344.

[136]   Pistollato F, Iglesias RC, Ruiz R, *et al*. Nutritional patterns associated with the maintenance of neurocognitive functions and the risk of dementia and Alzheimer's disease: A focus on human studies. Pharmacol Res 2018; 131: 32-43.
[http://dx.doi.org/10.1016/j.phrs.2018.03.012] [PMID: 29555333]

[137]   Dash S, Clarke G, Berk M, Jacka FN. The gut microbiome and diet in psychiatry: focus on depression. Curr Opin Psychiatry 2015; 28(1): 1-6.
[http://dx.doi.org/10.1097/YCO.0000000000000117] [PMID: 25415497]

[138]   Ezra-Nevo G, Henriques SF, Ribeiro C. The diet-microbiome tango: how nutrients lead the gut brain axis. Curr Opin Neurobiol 2020; 62: 122-32.
[http://dx.doi.org/10.1016/j.conb.2020.02.005] [PMID: 32199342]

[139]   Scheperjans F, Aho V, Pereira PA, *et al*. Gut microbiota are related to Parkinson's disease and clinical phenotype. Mov Disord 2015; 30(3): 350-8.
[http://dx.doi.org/10.1002/mds.26069] [PMID: 25476529]

[140]   Braak H, Müller CM, Rüb U, *et al*. Pathology associated with sporadic Parkinson's disease—where does it end?Parkinson's Disease and Related Disorders. Vienna: Springer 2006; pp. 89-97.
[http://dx.doi.org/10.1007/978-3-211-45295-0_15]

[141]   Sampson TR, Debelius JW, Thron T, *et al*. Gut microbiota regulate motor deficits and

neuroinflammation in a model of Parkinson's disease. Cell 2016; 167(6): 1469-1480.e12.
[http://dx.doi.org/10.1016/j.cell.2016.11.018] [PMID: 27912057]

[142] Savica R, Carlin JM, Grossardt BR, *et al.* Medical records documentation of constipation preceding Parkinson disease: A case-control study. Neurology 2009; 73(21): 1752-8.
[http://dx.doi.org/10.1212/WNL.0b013e3181c34af5] [PMID: 19933976]

[143] Forsyth CB, Shannon KM, Kordower JH, *et al.* Increased intestinal permeability correlates with sigmoid mucosa alpha-synuclein staining and endotoxin exposure markers in early Parkinson's disease. PLoS One 2011; 6(12): e28032.
[http://dx.doi.org/10.1371/journal.pone.0028032] [PMID: 22145021]

[144] Niehaus I, Lange JH. Endotoxin: is it an environmental factor in the cause of Parkinson's disease? Occup Environ Med 2003; 60(5): 378-8.
[http://dx.doi.org/10.1136/oem.60.5.378] [PMID: 12709528]

[145] Guo C, Wang P, Zhong ML, *et al.* Deferoxamine inhibits iron induced hippocampal tau phosphorylation in the Alzheimer transgenic mouse brain. Neurochem Int 2013; 62(2): 165-72.
[http://dx.doi.org/10.1016/j.neuint.2012.12.005] [PMID: 23262393]

[146] Dobbs RJ, Charlett A, Purkiss AG, Dobbs SM, Weller C, Peterson DW. Association of circulating TNF-α and IL-6 with ageing and parkinsonism. Acta Neurol Scand 1999; 100(1): 34-41.
[http://dx.doi.org/10.1111/j.1600-0404.1999.tb00721.x] [PMID: 10416510]

[147] Reale M, Iarlori C, Thomas A, *et al.* Peripheral cytokines profile in Parkinson's disease. Brain Behav Immun 2009; 23(1): 55-63.
[http://dx.doi.org/10.1016/j.bbi.2008.07.003] [PMID: 18678243]

[148] Sui YT, Bullock KM, Erickson MA, Zhang J, Banks WA. Alpha synuclein is transported into and out of the brain by the blood-brain barrier. Peptides 2014; 62: 197-202.
[http://dx.doi.org/10.1016/j.peptides.2014.09.018] [PMID: 25278492]

[149] Fasano A, Bove F, Gabrielli M, *et al.* The role of small intestinal bacterial overgrowth in Parkinson's disease. Mov Disord 2013; 28(9): 1241-9.
[http://dx.doi.org/10.1002/mds.25522] [PMID: 23712625]

[150] Tan AH, Mahadeva S, Thalha AM, *et al.* Small intestinal bacterial overgrowth in Parkinson's disease. Parkinsonism Relat Disord 2014; 20(5): 535-40.
[http://dx.doi.org/10.1016/j.parkreldis.2014.02.019] [PMID: 24637123]

[151] Keshavarzian A, Green SJ, Engen PA, *et al.* Colonic bacterial composition in Parkinson's disease. Mov Disord 2015; 30(10): 1351-60.
[http://dx.doi.org/10.1002/mds.26307] [PMID: 26179554]

[152] Hasegawa S, Goto S, Tsuji H, *et al.* Intestinal dysbiosis and lowered serum lipopolysaccharide-binding protein in Parkinson's disease. PLoS One 2015; 10(11): e0142164.
[http://dx.doi.org/10.1371/journal.pone.0142164] [PMID: 26539989]

[153] Li W, Wu X, Hu X, *et al.* Structural changes of gut microbiota in Parkinson's disease and its correlation with clinical features. Sci China Life Sci 2017; 60(11): 1223-33.
[http://dx.doi.org/10.1007/s11427-016-9001-4] [PMID: 28536926]

[154] Varankovich NV, Nickerson MT, Korber DR. Probiotic-based strategies for therapeutic and prophylactic use against multiple gastrointestinal diseases. Front Microbiol 2015; 6: 685.
[http://dx.doi.org/10.3389/fmicb.2015.00685] [PMID: 26236287]

[155] Gazerani P. Probiotics for Parkinson's disease. Int J Mol Sci 2019; 20(17): 4121.
[http://dx.doi.org/10.3390/ijms20174121] [PMID: 31450864]

[156] Mischley LK, Lau RC, Bennett RD. Role of diet and nutritional supplements in Parkinson's disease progression. Oxid Med Cell Longev 2017; 2017: 6405278.
[http://dx.doi.org/10.1155/2017/6405278] [PMID: 29081890]

[157] Mattson MP, Arumugam TV. Hallmarks of brain aging: adaptive and pathological modification by metabolic states. Cell Metab 2018; 27(6): 1176-99.
[http://dx.doi.org/10.1016/j.cmet.2018.05.011] [PMID: 29874566]

[158] Vauzour D. Dietary polyphenols as modulators of brain functions: biological actions and molecular mechanisms underpinning their beneficial effects. Oxid Med Cell Longev 2012; 2012: 914273.
[http://dx.doi.org/10.1155/2012/914273] [PMID: 22701758]

[159] UYAR GÖ, Yildiran H. A nutritional approach to microbiota in Parkinson's disease. Bioscience of microbiota, food and health 2019; 19-002.

CHAPTER 6

Therapeutic Efficacy of Mushroom in Neurodegenerative Diseases

Ankita Kushwaha[1,*], Vivek K. Chaturvedi[2], Sachchida Nand Rai[2], Sanjay C. Masih[3] and M. P. Singh[2]

[1] *Centre of Biophysics, Ewing Christian College, Prayagraj-211003, India*

[2] *Centre of Biotechnology, University of Allahabad, Prayagraj-211002, India*

[3] *Department of Zoology, Ewing Christian College, Prayagraj-211003, India*

Abstract: Mushrooms are used not only for culinary purposes, but also for the treatment of various chronic diseases. It shows vital therapeutic activity in several neurodegenerative disorders such as, Alzheimer's and Parkinson's diseases. These diseases are non-communicable as well as age-related. Currently, no drug therapy is available to treat such neurodegenerative disorders; instead, it is best to delay progression of these diseases. Accumulated evidence has suggested that culinary or medicinal mushrooms may play a significant role in the prevention of these disorders, as mentioned earlier, and dementia. Therefore, daily consumption of mushrooms in the diet may improve memory and cognitive functions, including mushrooms such as, *Hericium Erinaceus, Ganoderma lucidium, Pleurotus giganteus, Dictyophora indusiata, Sarcodon scabrosus, Antrodia camphorata Termitomyces albuminosus, Paxillus panuoides, Mycoleptodonoides aitchisonii, Lignosus rhinocerotis,* and numerous other species. These mushrooms show potent antioxidative, anti-inflammatory, and memory-enhancing activities. This chapter deals with the therapeutic activity of mushrooms and their bioactive components for different neurodegenerative diseases. Thus, mushrooms can be considered supportive and promising candidates for treating or preventing neurodegenerative diseases.

Keywords: Anti-inflammatory, Antioxidant, Culinary mushroom, Neurodegeneration, Neuritogenic, Neuroprotection, Neurotrophic.

INTRODUCTION

Neurodegenerative diseases (NDs) are predominantly increasing, attributable, in part, to extensions of a lifetime, which in turn, poses a significant danger to human health. Unfortunately, many of these diseases will increase as the world population ages and is expected to double by 2050 [1]. There is no treatment for

* **Corresponding author Ankita Kushwaha:** Centre of Biophysics, Ewing Christian College, Prayagraj-211003, India; Tel: +91 8299620239; E-mail:eshacompact15@gmail.com

NDs that prevent progressive degeneration of the nerve cells [2]. Alzheimer's disease (AD), Parkinson's disease (PD), Huntington's disease (HD), amyotrophic lateral sclerosis (ALS), frontotemporal dementia, and spinocerebellar ataxias are examples of neurodegenerative diseases affecting millions of people all around the world. Such conditions are different in terms of their pathophysiology, such as loss of memory, cognitive (called dementia), and other abilities that affect the daily activity of an individual (called ataxias) [3, 4]. Currently, therapeutic neuroprotective approaches are explored to target molecular mechanisms that give rise to neurodegenerative diseases. Oxidative stress and mitochondrial dysfunction are the two factors that play a crucial role in the development of neurodegenerative diseases. The generation of toxic reactive oxygen species (ROS) interrupts the normal cellular metabolism, which causes extensive damage to the cells and tissues, and also causes neuronal cell death, which ultimately leads to oxidative stress [5]. Disruption in the mitochondrial functions induces glutamate-induced neuronal neurotoxicity, as mitochondria help to regulate the cell-death process. Therefore, oxidative stress and a mutation in the mitochondrial DNA cause neurodegenerative disease [6]. This chapter focuses on alternative drug therapy by daily intake of various culinary and medicinal mushrooms in the diet, leading to a delay in the development of neurodegenerative diseases. Mushrooms are macrofungi placed in the subkingdom of Dikarya, phyla Basidiomycota, and Ascomycota, whereas in class of Agaricomycetes and Pezizomycetes, respectively, they are also ubiquitous. Mushrooms are widely known not only for their nutritional properties, but also for their medicinal properties worldwide [7]. Mushrooms have a lot of beneficial compounds, such as, free radical scavengers, anti-apoptotic factors, and stimulators of nerve growth factors (NGF). These compounds directly exert positive effects on the brain. Thus, they serve as a "neuro-nutraceuticals" that protect the neuronal cell both *in vivo* and *in vitro*.

NEUROPROTECTIVE EFFECTS OF MUSHROOMS AGAINST NEURODEGENERATIVE DISEASE

Currently, the number of mushrooms estimated on the earth's surface are 150,000–160,000, although only 10% of the species are known [8]. They are known for their medicinal properties, excessive vitamins, free radical scavenger's antioxidant food supplements, and nitrosative stress to degenerate the metabolic processes [9]. The present section deals with the neuroprotective effects of mushrooms and their molecular mechanism against neurodegenerative disease. *Hericium Erinaceus is* also known as Lion's mushroom or Pom Pom mushroom, and it has been eaten for several years in Asian countries. Hericenones, a benzyl alcohol derivative (A-H), and Erinacines (A-K and P-Q) are two bioactive 2°

metabolites from fruiting bodies(basidiocarps) and mycelium, respectively [10] (Table **1**). Hericenones A and B were reported earlier, in the 1990s, but no neurite outgrowth NGF activity was reported. Moreover, in Hericenones C-H, *in vitro,* the biosynthesis of NGF was documented [11], whereas, Hericenones I and J and 3-hyrdoxyhericenone F, were reported in 2008. However, only 3-hydroxyhericenone showed a protective function against the ER stress-dependent neuro2a cell death. Erinacines A-C was an intense stimulation of the NGF activity [12]. The bioactive compound Dilinoleoyl Phoshatidylethanolamine (DLPE) from basidiocarps of *H. erinaceum* reduced the ER stress-dependent neuro2a cell death. Nagai *et al.* [13], also reported that 100 ng/ml DLPE from *H.erinaceum* significantly reduced the cell viability of Neuro2a cells, when treated with tunicamycin; thus, cell death occurred *via* the protein kinase C pathway (PKC). When the oral administration of *H. erinaceum* in the rat's brain was studied, significantly increased lipoxin A4 (LXA4) protein (anti-inflammatory properties) was found in the cortex and hippocampus region of the brain. Moreover, LXA4 upregulation was associated with the increased expression of cytoprotective proteins such as heat shock protein 70 (Hsp70), hemeoxygenase-11(HO-1), and thioredoxin (TRX), as investigated by Salinaro *et al.* [14]. *T. albuminous* is known as Termite Mushroom or "Ji Zong" mushroom in Chinese and consumed in countries such as China, Japan, and Chile. Cerebrosides and termitomycesphins (A-H) were bioactive compounds, extracted from the dried basidiocarps of *T.albuminosus.* When 10 µg/ml of termitomycesphins (A-D) was treated for six days, it induced neurite outgrowth of PC12 cell by 20% and 25%, whereas, 1 µM termitomycesphins G and H did the same in only for 48 hours [15, 16]. Choi *et al.* also investigated termitomycamide (A-E), which was a fatty acid amide. When 0.1 µg/ml of termitomycamide B and E was treated, it significantly reduced ER stress-dependent Neuro2a cell death by 20%. *M. aitchisonii,* also known as "Bunaharitake" mushroom in Japan, is cultivated on dead broadleaf trees in Asia and is well known for its beneficial effects of treating persistent diseases. There was *in-vitro* investigation of 0.6 µM of 5-hydroxy-4-(1-hydroxyeth-l)-3-methylfuran-2(5H)-one and 5-phenylpentane-1,3,4-triol; they were treated for 24 hours. It was observed that the two above-mentioned bioactive compounds protected against Endoplasmic Reticulum (ER) stress-dependent Neuro2a cell death [17]. However, when 0.1 µg/ml of other bioactive compounds such as 3-(hydroxymethyl)-4-methylfuran-2(5H)-one and (3R,4S, 1'R)-3-(1'-hydr-xy-ethyl)-4-methyldihydrofuran-2(3H)-one were treated for 24 hours, they showed significantly reduced tunicamycin-induced neuronal cell death, which signified their protective effects against ER stress-dependent Neuro2a cell death [18].

Table 1. Mushrooms that have potential neuroprotective effects against neurodegenerative disorders.

Name of the Mushroom	Bioactive 2° Metabolites	Neuroprotective Mechanism	Therapeutic Application	References
Hericium erinaceum	(i) Erinacines (A-I and P-Q),	Induces NGF(Neurotrophic) Synthesis	Prevents AD, PD, HD and ALS	[37, 38]
	(ii) Hericenones (A-H)			
	(iii) 4-chloro-3,5-dimethoxybenzoic methyl ester	Exhibits potent neuroprotective activity		[39]
	(iv) Lipoxin A4	Decreased ER stress Increases mitochondrial function		[14]
Termitomyces albuminosus	Saponins	Anti-inflammatory activity	Prevents neuronal cell death induced by neurodegenerative disease	[40, 41]
	Chitin-glucan complex	Promotes neurite growth		[42]
	Termitomycesphins (B and E)	Reduces oxidative stress		[43]
Mycoleptodonoides aitchisonii	3-Hydroxymethyl-4-methylfuran-25H-one, 3R,4S,1'R-3--'-hydroxyethyl-4methyldihydrofuran-23H-one, 5-hydroxy-4-1-hydroxyethyl-3-e-methylfuran-25H-one, 5-phenylpentane-1,3,4-triol	Protects Neuro 2a cells from ER stress, Increases NGF synthesis	Prevents neuronal cell death induced by neurodegenerative disease	[44]
Daldinia concentrica	Volatile organic compounds	Neuroprotective activity	Prevention of Neurodegeneration	[45]
Dictyophora indusiata	Dictyophorines A and B	Induces NGF		[45]
	Dictyoquinazol A-C	Protects cortical neurons from excitotoxicity		[46, 47]
Cortinarius infractus	6-Hydroxyinfractine, infractopicrine	Decreases AChE activity, reduces Aβ Aggregation	Prevention of AD	[48]
Antrodia camphorata	Diterpenes	Antioxidant, Anti-inflammatory, Antipoptotic and reduces tau protein	Prevention of AD	[25]
Pleurotus giganteus	Uridine	Neurotrophic, Neuritogenic:MEK/ERK1/2 and P13K pathway	Prevents neuronal cell death induced by neurodegenerative disorders	[38]
	Linoleic acid	Neuritogenesis in PC 12 cells		[38]
P. cornucopiae	Ergothioneine	Induces neuronal precursor double cortin cells: increases neurogenesis in mice		[49]
Ganoderma lucidum	Ganodermaside A – D	Regulates aging-related gene *UTH1* in yeast and prolongs the lifespan	Prevention of AD	[29]
	Ganolucidic acid A, S1,TQ, Methyl ganoderic acid A and B, Ganodermatriol, 7-oxo-ganoderic acid Z, 4,4,14-trimethyl-5-chol-7,9(11)-dien-3--oxo-24-oic acid	Have BDNF-like neurotrophic activities		[50]
Ganoderma neo-japonicum		Induced NGF Synthesis Antioxidant activity	Prevents the neuronal cell death induced by neurodegenerative disorders	[30]
Lignosus rhinocerotis	Polysaccharides	Mimic NGF activity:MEK/ERK1/2 pathway	Prevents neuronal cell death induced by neurodegenerative disorders	[51]

(Table 1) cont.....

Name of the Mushroom	Bioactive 2° Metabolites	Neuroprotective Mechanism	Therapeutic Application	References
Sarcodon scabrosus	Sarcodonin A,G, M	Induces neurite outgrowth	Prevents neuronal cell death induced by neurodegenerative disorders	[52]
	Scabronine H,J,K and19-O-acetylsarcodonin G	Induces NGF synthesis		[32]
Paxillus panuoides	Leucomentin-1, Leucomentin-2 and Leucomentin-3	Antioxidant activity	Prevention of AD-like Neurogenerative disorder	[34]
Coriolus versicolor	LXA4	Anti-inflammatory	Prevention of AD and other neurodegenerative disorders	[53]
Cordyceps militaris	Cordycepin (1), Adenosine	Increased BDNF expression	Prevents AD	[36]
Grifola frondosa	Lysophosphatidylethanolamine	Neuritogenic effect	Prevents neuronal cell death induced by neurodegenerative disorders	[54]
Cortinarius brunneus	β-Carboline	Inhibits AChE activity	Prevents AD	[24]

Daldinia concentrica has several common names, such as King Alfred's cake, cramp balls, and coal fungus. Two neuroprotective compounds, 1-3,4,--Trimethoxyphenyl ethanol and caruilignan, have been identified in the fruiting body of *D. concentrica*. Both compounds have neuroprotective effects against iron-induced neurodegeneration in mouse cortical neurons [19]. *Dictyophora indusiata*, also known as Bamboo Mushroom or Veiled lady Mushroom, possesses medicinal and culinary properties and is found in Asian countries [20]. Dictyophorines A and B are the potent bioactive compounds isolated from the mushroom described above and can considerably induce the Nerve Growth Factor (NGF) in astroglial cells [21]. In contrast, Dictyoquinazol A-C is identified in *D. indusiata*, and it has been found that 5 μM of Dictyoquinazol A, B, and C treatment can significantly protect the cortical neuron cell death from excitotoxicity [22].

Mushrooms of the *genus Cortinarius* has two species, *C. infractus* and *C. brunneins, which* are studied for their peculiar enzyme acetyl cholinesterase(AChE)-inhibiting abilities. The two bioactive compounds like infractopicrin and 10-hydroxy-infractopicrin were isolated from the extracts of the fruiting bodies of *C. Infractus and* were studied for their potent AChE-inhibiting ability along with the Aβ-peptide self-aggregation, thus preventing AD disease [23]. The fruiting bodies of *C. Brunneins* exhibits the β-Carboline bioactive compound, which is called Brunneins. Brunneins-A showed evidence of very low AChE-inhibiting activity and no cytotoxicity [24]. *Antrodia camphorate.*, a traditional medicinal mushroom commonly found in Taiwan and Asia, is used to treat various inflammatory disorders. Diterpenes were isolated from the basidiocarps of *A. camphorata* and demonstrated their neuroprotective properties against the Aβ-induced neurotoxicity in the cortical neurons, as shown in Table **1** [25]. *Pleurotus giganteus*, one of the edible mushrooms, exhibited high Uridine content (1.8/100 g) in the extract of mushroom. The mushroom extract revealed a neurite outgrowth by mimicking the NGF activity *via* the MEK/ERK1/2 and P13K signaling pathway (Phan *et al.*, 2015). In another study, Uridine, the bioactive compound was responsible for the neuronal outgrowth (Chia Wei 2015). Golden oyster mushroom called *P. cornucopiae is* known as a health-promoting body energy restorer and it exhibits ergothioneine as a bioactive compound that acts as a potent antioxidant. It functioned as a neuroprotectant [26]. *Ganoderma lucidum,* a well-known medicinal mushroom used in the treatment of chronic diseases such as cancers and neurodegenerative disorders, is also prominent for its neuroprotective properties for thousands of years in Asia (especially in the eastern part) with the common name of "Ling Zhi" in Chinese. Numerous polysaccharides such as steroids, fatty acids, nucleotides, triterpenoid sterols possessed immunomodulatory, strong antioxidants, and other beneficial compounds that were found to be protective against cerebral ischemic injury [27]. These bioactive secondary metabolites possessed the age-regulated gene that elongated the lifespan of yeast by increasing the Nerve Growth Factor (NGF) and Brain-derived neurotrophic factor (BDNF); thereby, neuronal cell death was prevented and protected [28, 29]. *Ganoderma neo-japonicum was* studied for its NGF that induced neurite outgrowth [30].

The common name of Lignosus rhinocerus is "Tiger's milk mushroom." The mushroom has polysaccharides that are knownto significantly induce neuritogenesis without NGF production in PC 12 cells by mimicking the MEK/ERK1/2 pathway [31]. *Sarcodon scabrosus* is a bitter-non-edible mushroom found in the coniferous forests of Japan. The alkaloid, scabronine, is isolated from the fruiting bodies of *S. scabrosus.* Bioactive compounds, erinacines, are separated from the fruiting bodies of *H. erinaceus.* Sarcodonines A and G are the strong inducers of NGF biosynthesis and induce neurite outgrowth, respectively [32, 33]. *Paxilluspanuoides* help in the way of being potent antioxidants. The bioactive compounds, such as leucomentin (p-terphenyl compound), are capable of neuroprotective activity, and their mechanism has been investigated. In addition, they are potent lipid peroxidation inhibitors and induce $H2O2$ neurotoxicity [34]. Sabaratnam *et al.* [35], after an investigation has found that the bioactive compound cordycepin is isolated from *Cordyceps militaris*. In general, neurodegeneration or brain-derived neurotrophic factor (BDNF) conditions are significantly reduced in the hippocampus tissue of the brain. However, it is the opposite in the case of cordycepin, as it shows increased BDNF expression in cordycepin-administrated mice and decreased amount of inflammation markers such as (NF)-κBP65 TNF-α and IL-6, which represented the presence of antidepressants [36]. Fig. (**1**) shows the neuroprotective activity of several mushrooms in neurodegenerative diseases.

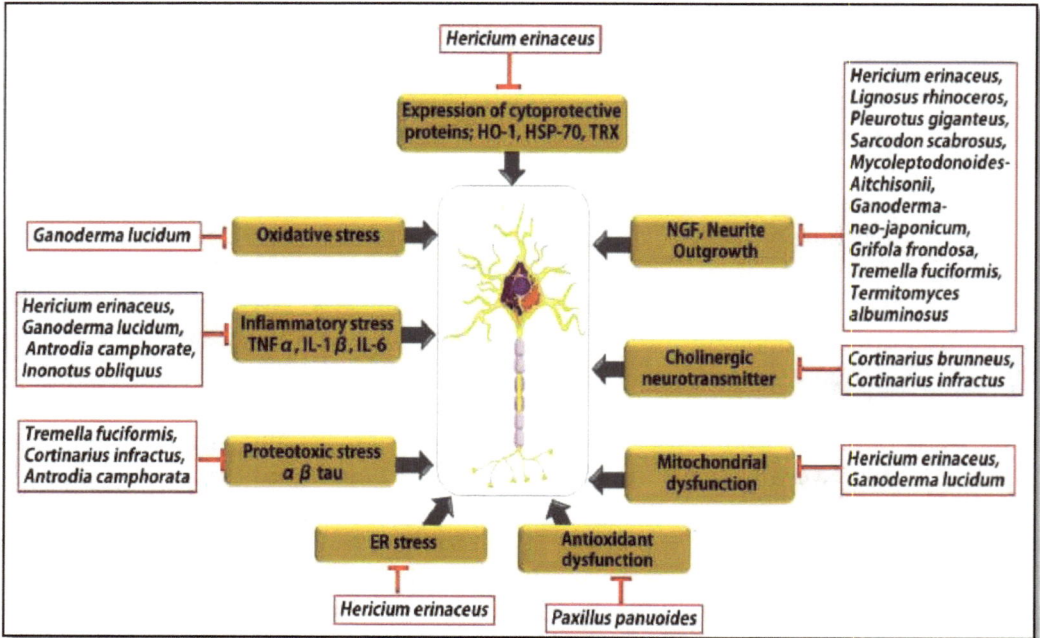

Fig. (1). Neuroprotective mechanism of various mushrooms in neurodegenerative disease.

ADVANTAGEOUS EFFECTS OF MUSHROOMS AGAINST NDS

Role of Mushrooms in Alzheimer's Disease (AD)

Alzheimer's disease is an incurable neurodegenerative cognitive disorder. It is the most widely recognised age-related dementia [55]. Older people are mainly affected by Alzheimer's disorder. The symptoms initially affect short-term memory. Impairment of mood, functional activities and behavior of an individual, such as, impaired speech, language perception, memory, communication, execution of motor functions like reasoning and judgement and finally, progressive changes in neuronal dysfunction (dementia) in the central nervous system take place [56]. The principal hallmarks of AD are, (i) the presence of extracellular deposition of beta-amyloid (Aβ) peptide plaques and (ii) the protein accumulation of intracellular hyperphosphorylated tau proteins, known as neurofibrillary tangles, in the subcortical regions of the brain [57, 58]. The most toxic variant of the amyloid protein is Aβ1–42, and it can aggregate into dimers, oligomers, and fibrils. Amyloidosis inevitably causes oxidative stress, inflammatory processes and acceptance of neuronal apoptosis in the cerebrums of those with AD. The primary causative factors for AD pathogenesis are tentative. Still, brain aging, oxidative stress, distortion of mitochondrial activity, inflammation, and energy imbalance lead to the disruption of neuronal activity (autophagic/apoptotic cell death) and CNS functioning [59]. The current drug therapy for neurodegenerative diseases is ineffective, with many facet effects, and it best provides a short-term delay in the development of AD. Additionally, the drug improvement pipeline is drying up, and the number of innovative drug therapies achieved in the marketplace has lagged behind the growing want for such drug therapy [60]. Daily consumption of mushrooms with potent antioxidant activity and useful immunomodulating properties is the alternative approach to mitigate AD.

Cilaerdzic *et al.* [61], reported that the *Pleurotusos treatus* and *Laetiporus sulphurous* possess a higher reducing power, an efficient scavenger of DPPH and an ABTS radical. It also includes a healthy Fe^{3+} reducing activity and has efficient acetylcholinesterase- and tyrosinase-inhibiting properties. These species, owing to the significant antioxidant and neurodegenerative capacity; can therefore, be suggested as possible agents for Alzheimer's and Parkinson's Diseases. Wong *et al.* [62], investigated that the aqueous extracts of mycelium of *H. erinaceus* stimulate the neurite outgrowth *in vitro* from the neural cell-cultured hybrid clone NG108-15, as it contains neuroactive compounds that help in the treatment of neurological disorders. The extract also induces the neurite outgrowth, which stimulates the NGF activity when it amalgamates with the NG108-15 cell line studied by Lai *et al.* [63]. Mori *et al.* [64], studied the effects of ethanol extracts of four different mushrooms, namely *H. erinaceus*, *P. eryngii*, *G. frondosa* and *A. blazei* on nerve growth factor (NGF) mRNA gene expression and proteins, in 1321N1 human astrocytoma cells. Among all four mushrooms, only *H. erinaceus* promotes the NGF inducing activity and enhances neurite outgrowth of PC12 pheochromocytoma cells *via* c-Jun N-terminal kinase activity. Mori *et al.* [65], suggested that *H. erinaceus* effectively improves mild cognitive impairment, without any adverse effects.

They documented that the subjects of the *H. erinaceus* group took four 250 mg tablets containing 96% of dry powder of *H. erinaceus,* three times a day, for 16 weeks. After termination of the intake, the subjects were observed for the next four weeks. At weeks 8, 12 and 16 of the trial, the *H. erinaceus* group showed significantly increased scores on the cognitive function scale compared with the placebo group. The *H. erinaceus* group's scores increased with the duration of intake. Both *in-vitro* and *in vivo* studies showed no adverse effect of *H. erinaceus.* According to Mori *et al.* [66], a diet containing *H. erinaceus* powder prevented the spatial short-term and visual recognition memory (cognitive dysfunction) loss induced by intra-cerebroventricular administration of the Aβ (25-35) peptide.

According to Kuo *et al.* [67], a human pilot experiment evaluated the anxiety and sleep quality effects of four weeks of administration of Amyloban® 3399 (which was obtained from *H. erinaceus)* in undergraduate students of Japan. Each student consumed six tablets per day. The results revealed an increase in salivary free-MHPG, which corresponded to an improvement in anxiety and sleep quality. Therefore, they concluded that one of the possible effects of Amyloban® 3399 was to balance out the mind and body. Karami *et al.* [68], investigated that cholinergic neuron loss affected the NGF activity in patients with Alzheimer's disease. Choline acetyltransferase (ChAT) and acetyltransferase were cholinergic biomarkers found in the cerebrospinal fluid (CSF). ChAT showed a higher correlation with cognition in comparison to acetyltransferase. Various compounds from mushrooms target ChAT for neuroprotective effect [69]. Nagano *et al.* [70], investigated the clinical impact of *H. erinaceus* on menopause, depression and sleep quality. Thirty female subjects of Japan were randomly selected in the placebo group and the *H. erinaceus* (HE) group and requested to take these mushroom cookies for four weeks. Therefore, we could state that HE possessed potent neuroprotective activity, as indicated by the AD mouse model and *in vitro* study [71]. The potential neuroprotective properties of *Amanita caesarea (*AC) were shown by Li *et al.*, in the L-Glutamic acid (L-Glu)-induced HT22 apoptotic cell model and in the AlCl3 and D-galactose(D-gal)-induced experimental AD mice model [72]. Jeong *et al.* [73], reported that the protoplast of *P. eryngii* was mutagenised using 4-nitroquinoleneoxide in order to improve the medicinal qualities. *In vitro* studies were carried out by applying aqueous extracts of *P. Eryngii* to the human hepatoma cell line HepG2. The *P.eryngii* mutant, NQ2A-12, was shown to have an enhanced expression of Pin 1 in HepG2, in a mouse brain tissue. Thus, this mushroom had the potential to protect against AD. A Mediterranean diet (MD) containing oyster mushrooms such as *P. ostreatus* and *P. eryngii* had been proposed, as a reduced risk for AD and cognitive impairment [74, 75]. Chong *et al.* [76], documented a clinical study on patients (62 females and 15 males overweight or obese participants) with depression, anxiety and sleep disorder, in which the beneficial effects of eight weeks of oral administration of capsules containing *H. erinaceus* extract (water and ethanolic) (1200 mg/capsule, 3 capsules/day) had been examined. The supplementation of *H. erinaceus* improved mood disorders of a depressive-anxious nature and the quality of the nocturnal rest. *H. erinaceus* increased circulating pro-BDNF levels, without any significant change in the BDNF circulating levels.

Bennett *et al.* [77], showed that the vitamin D2-enriched button mushroom (*Agaricus bisporus*) improves memory in wild-type and APPswe/PS1dE9 transgenic mice. Two-month-old wild-type (B6C3) and AD transgenic (APPSwe/PS1dE9) mice were fed a diet either deficient in Vitamin D2 or a diet that was supplemented with VDM for 7 months. They compared them with mice fed on the control diet, VDM-fed, wild-type and AD transgenic mice displayed improved learning and memory, had significantly reduced amyloid plaque load and glial fibrillary acidic protein and elevated interleukin-10 in the brain. The results suggested that VDM might provide a dietary source of Vitamin D2 and other bioactive compounds for preventing memory-impairment in dementia. Phan *et al.* [78], investigated that the aqueous extracts of *Pleurotus giganteus* were found to stimulate neurite outgrowth *in vitro*. Uridine of 100 μM was found to enhance the percentage of neurite-bearing cells of differentiating neuroblastoma (N2a) cells. Uridine, which was present in *P. giganteus*, also increased the phosphorylation of extracellular-signal-regulated kinases (ERKs) and protein kinase B (Akt). Also, phosphorylation of the mammalian target of rapamycin (mTOR) was increased. MEK/ERK and PI3K-Akt-mTOR further induced phosphorylation of the cAMP-response of the element-binding protein (CREB) and expression of the growth-associated protein 43 (GAP43), which all together promoted neurite outgrowth of N2a cells. Linoleic acid played a vital role in neuritogenesis in rat PC12 cells [79].

Non-genetic sporadic AD models were developed by single-time intracerebroventricular (ICV) infusion of streptozotocin (STZ) in rats. It exhibited protein aggregation (amyloidosis) in rats and dementia hippocampus, probably resulting from increased cellular mitochondrial fragmentation, and functional aberrations [80]. Pre-administration with Ganoderma lucidum spore (GLS) in ICV STZ model rats exhibited an improvement in biochemical parameters (Zhou *et al.*, 2012). Choi *et al.*, (2015) showed the anti-amnesic effect of fermented *G. Lucidum* water extracts (GW) by lactic acid bacteria on scopolamine. They showed that *G. Lucidum* was useful in enhancing learning, memory and cognitive function *via* cholinergic dysfunction. In the mouse model of AD, Huang *et al.* [81], reported that oral administration of the polysaccharides and water extract from *G. lucidum* promoted neural progenitor cell (NPC) proliferation, which enhanced neurogenesis.

P. ostreatus, stored at various temperatures, also showed antioxidative properties and might be utilised to treat neurodegenerative diseases [82]. Lee *et al.* [83], showed that the ethanol extract of *G. lucidium* had neuroprotective effects in murine hippocampal HT22 cells, which would be potential pharmaceutical products for brain disorder induced by neuronal damage and oxidative stress. The oral administration of a *Coriolus versicolor* (Cv) biomass was given for 30 days and the expression of LXA4 was measured in the rats' brains. Thus, Cv prevented progressive neurodegeneration, which was also shown by clinical trials [84].

Therefore, Cv possessed vital neuroprotective activity for neurodegenerative diseases [85].

Mushrooms Having Beneficial Effects Against Parkinson's Pisease (PD)

Parkinson's disease (PD) is a chronic and age-related neurodegenerative disease, first described by James Parkinson as 'Shaking Palsy,' in the 1800s. PD is caused by the consistent loss of dopaminergic neurons within the midbrain's substantia nigra pars compacta (SNpc). Furthermore, it develops abnormal motor function, including resting tremor, postural instability and bradykinesia. The PD patients also suffer from non-motor complications such as sleep, anxiety, depression and cognitive functions. Certain autonomic and sensory functions are the non-motor complications that the PD patients show [86]. PD is the second-most common neurodegenerative disorder after AD. The neuropathological features of PD are the accumulation of intracellular protein aggregates, Lewy bodies and Lewy neuritis, responsible for the progression of this disease. The most common synucleinopathy, in addition to cerebral amyloid disorder, is categorised under PD [86]. Oxidative stress, nitrosative stress, mitochondrial dysfunction and lysosomal dysfunction are the major factors behind PD [87]. In a population that is more than 65 and 85 years of age, epidemiologically, it has been assessed that about 1–2% and 3–5%, respectively, suffer from PD [88]. Certain evidence has suggested that regular intake of antioxidants, L-type calcium channel antagonists and non-steroidal anti-inflammatory drugs (NSAIDs) might be utilised to prevent the progression of this disease [89, 90].

Zhu *et al.* [91] have shown that the oil from *G.lucidium*s pores has a neuroprotective effect for preventing the entry of dopaminergic neurons within SNpc, in mouse models of PD. Gobi *et al.* [92], investigated the anti-PD effects of the aqueous extract of *Agaricus blazei* mushroom (5 µg/100 µl) against the rotenone-induced (100 nM) *in vitro* model of PD. The treatment with *A. blazei* extracts considerably suppressed the rotenone-induced neurotoxicity, by attenuating the levels of ROS, the mitochondrial transmembrane potential (MMP), apoptosis and significantly enhanced cell viability. Thus, *A. blazei* may be a promising drug for the treatment of PD. In another study, the treatment was with methanolic extracts of *A. blazei* (20 g/200 ml) against the rotenone-induced PD in male albino mice. Rotenone treatment significantly enhanced the motor impairments, neuroinflammation and oxidative stress. On the other hand, oral administration of the methanol extract of *A. blazei* attenuated the indices described above. Thus, it is concluded that *A. blazei* is a promising drug candidate for PD treatment [93].

CONCLUSIONS

In this book chapter, we have concluded that mushrooms and their different bioactive components play a vital role in neurodegenerative diseases. The mushroom and its bioactive components exhibit potent antioxidative and anti-inflammatory activities. They offer significant neuroprotective activity in different toxin-induced models of Parkinson's and Alzheimer's disease. Mushrooms also promote the activity of other growth factors. The bioactive components of mushrooms are of considerable importance, as they display various functions in neurodegenerative diseases. Further studies will be needed to characterise the additional features of mushrooms and their vital therapeutic activities.

CONSENT FOR PUBLICATION

Not applicable.

CONFLICT OF INTEREST

The authors declare no conflict of interest, financial or otherwise.

ACKNOWLEDGEMENTS

Authors would like to acknowledge UGC Dr. D.S. Kothari Postdoctoral scheme for awarding the fellowship to Dr. Sachchida Nand Rai (Ref. No-F.4-2/2006 (BSR)/BL/19-20/0032).

REFERENCES

[1]   Marsh SE, Blurton-Jones M. Neural stem cell therapy for neurodegenerative disorders: The role of neurotrophic support. Neurochem Int 2017; 106: 94-100.
[http://dx.doi.org/10.1016/j.neuint.2017.02.006] [PMID: 28219641]

[2]   Heemels MT. Neurodegenerative diseases. Nature. 2016 Nov 10;539(7628):179.
[http://dx.doi.org/10.1038/539179a] [PMID: 27830810]

[3]   Canter RG, Penney J, Tsai LH. The road to restoring neural circuits for the treatment of Alzheimer's disease. Nature 2016; 539(7628): 187-96.
[http://dx.doi.org/10.1038/nature20412] [PMID: 27830780]

[4]   Taylor JP, Brown RH Jr, Cleveland DW. Decoding ALS: from genes to mechanism. Nature 2016; 539(7628): 197-206.
[http://dx.doi.org/10.1038/nature20413] [PMID: 27830784]

[5]   Barnham KJ, Masters CL, Bush AI. Neurodegenerative diseases and oxidative stress. Nat Rev Drug Discov 2004; 3(3): 205-14.
[http://dx.doi.org/10.1038/nrd1330] [PMID: 15031734]

[6]   Lin MT, Beal MF. Mitochondrial dysfunction and oxidative stress in neurodegenerative diseases. Nature 2006; 443(7113): 787-95.
[http://dx.doi.org/10.1038/nature05292] [PMID: 17051205]

[7]   Journal O, Xu T, Beelman RB. The bioactive com-pounds in medicinal mushrooms have potential

protective effects against neu-rodegenerative diseases. Adv Food TechnolNutr Sci Open J 2015; 1(2): 62-6.
[http://dx.doi.org/10.17140/AFTNSOJ-1-110]

[8]   Wasser SP. Medicinal mushroom science: Current perspectives, advances, evidences, and challenges. Biomed J 2014; 37(6): 345-56.
[http://dx.doi.org/10.4103/2319-4170.138318] [PMID: 25179726]

[9]   Abdullah N, Ismail SM, Aminudin N, Shuib AS, Lau BF. Evaluation of selected culinary-medicinal mushrooms for antioxidant and ACE inhibitory activities. Evid Based Complement Alternat Med 2012; 2012: 464238.
[http://dx.doi.org/10.1155/2012/464238] [PMID: 21716693]

[10]   Williams RJ, Mohanakumar KP, Beart PM. Neuro-nutraceuticals: The path to brain health via nourishment is not so distant. Neurochem Int. 2015 Oct;89:1-6. doi: 10.1016/j.neuint.2015.08.012. Epub 2015 Aug 21.
[PMID: 26303091]

[11]   Kawagishi H, Ando M, Sakamoto H, *et al.* Hericenones C, D and E, stimulators of nerve growth factor (NGF)-synthesis, from the mushroom *Hericiumerinaceum.* Tetrahedron Lett 1991; 32(35): 4561-4.
[http://dx.doi.org/10.1016/0040-4039(91)80039-9]

[12]   Kawagishi H, Ando M, Shinba K, *et al.* Chromans, hericenones F, G and Hfrom the mushroom *Hericiumerinaceum.* Phytochemistry 1992; 32(1): 175-8.
[http://dx.doi.org/10.1016/0031-9422(92)80127-Z]

[13]   Nagai K, Chiba A, Nishino T, Kubota T, Kawagishi H. Dilinoleoyl-phosphatidylethanolamine from *Hericium erinaceum* protects against ER stress-dependent Neuro2a cell death *via* protein kinase C pathway. J Nutr Biochem 2006; 17(8): 525-30.
[http://dx.doi.org/10.1016/j.jnutbio.2005.09.007] [PMID: 16426828]

[14]   Trovato Salinaro A, Pennisi M, Di Paola R, *et al.* Neuroinflammation and neurohormesis in the pathogenesis of Alzheimer's disease and Alzheimer-linked pathologies: modulation by nutritional mushrooms. Immun Ageing 2018; 15(1): 8.
[http://dx.doi.org/10.1186/s12979-017-0108-1] [PMID: 29456585]

[15]   Qi J, Ojika M, Sakagami Y. Termitomycesphins A–D, novel neuritogenic cerebrosides from the edible Chinese mushroom *Termitomycesalbuminosus.* Tetrahedron 2000; 56(32): 5835-41.
[http://dx.doi.org/10.1016/S0040-4020(00)00548-2]

[16]   Qu Y, Sun K, Gao L, *et al.* Termitomycesphins G and H, additional cerebrosides from the edible Chinese mushroom *Termitomyces albuminosus.* Biosci Biotechnol Biochem 2012; 76(4): 791-3.
[http://dx.doi.org/10.1271/bbb.110918] [PMID: 22484955]

[17]   Choi JH, Suzuki T, Okumura H, *et al.* Endoplasmic reticulum stress suppressive compounds from the edible mushroom *Mycoleptodonoides aitchisoni.* J Nat Prod 2014; 77(7): 1729-33.
[http://dx.doi.org/10.1021/np500075m] [PMID: 24988471]

[18]   Choi JH, Horikawa M, Okumura H, *et al.* Endoplasmic reticulum (ER) stress protecting compounds from the mushroom *Mycoleptodonoidesaitchisonii.* Tetrahedron 2009; 65(1): 221-4.
[http://dx.doi.org/10.1016/j.tet.2008.10.068]

[19]   Lee IK, Yun BS, Kim YH, Yoo ID. Two neuroprotective compounds from mushroom Daldiniaconcentrica. Journal of microbiology and biotechnology, 2002a, 12(4), 692-694.

[20]   Habtemariam S. The Chemistry, Pharmacology and Therapeutic Potential of the Edible Mushroom *Dictyophora indusiata* (*Vent* ex. *Pers.*) Fischer (Synn. *Phallus indusiatus*). Biomedicines 2019; 7(4): 98.
[http://dx.doi.org/10.3390/biomedicines7040098] [PMID: 31842442]

[21]   Kawagishi H, Ishiyama D, Mori H, *et al.* Dictyophorines A and B, two stimulators of NGF-synthesis from the mushroom Dictyophora indusiata. Phytochemistry 1997; 45(6): 1203-5.

[http://dx.doi.org/10.1016/S0031-9422(97)00144-1] [PMID: 9272967]

[22]   Lee IK, Yun BS, Han G, Cho DH, Kim YH, Yoo ID. Dictyoquinazols A, B, and C, new neuroprotective compounds from the mushroom Dictyophora indusiata. J Nat Prod. 2002 Dec;65(12):1769-72.
[http://dx.doi.org/10.1021/np020163w] [PMID: 12502311]

[23]   Geissler T, Brandt W, Porzel A, *et al.* Acetylcholinesterase inhibitors from the toadstool *Cortinarius infractus.* Bioorg Med Chem 2010; 18(6): 2173-7.
[http://dx.doi.org/10.1016/j.bmc.2010.01.074] [PMID: 20176490]

[24]   Teichert A, Schmidt J, Porzel A, Arnold N, Wessjohann L. Brunneins A-C, β-carboline alkaloids from *Cortinarius brunneus.* J Nat Prod 2007; 70(9): 1529-31.
[http://dx.doi.org/10.1021/np070259w] [PMID: 17854153]

[25]   Chen CC, Shiao YJ, Lin RD, *et al.* Neuroprotective diterpenes from the fruiting body of *Antrodia camphorata.* J Nat Prod 2006; 69(4): 689-91.
[http://dx.doi.org/10.1021/np0581263] [PMID: 16643055]

[26]   Phan CW, David P, Naidu M, *et al.* Therapeutic potential of culinary-medicinal mushrooms for the management of neurodegenerative diseases: diversity, metabolite, and mechanism. Crit Rev Biotechnol. 2015;35(3):355-68.
[http://dx.doi.org/10.3109/07388551.2014.887649] [PMID: 24654802]

[27]   Gokce EC, Kahveci R, Atanur OM, *et al.* Neuroprotective effects of *Ganoderma lucidum* polysaccharides against traumatic spinal cord injury in rats. Injury 2015; 46(11): 2146-55.
[http://dx.doi.org/10.1016/j.injury.2015.08.017] [PMID: 26298021]

[28]   Weng Y, Lu J, Xiang L, *et al.* Ganodermasides C and D, two new anti-aging ergosterols from spores of the medicinal mushroom *Ganoderma lucidum.* Biosci Biotechnol Biochem 2011; 75(4): 800-3.
[http://dx.doi.org/10.1271/bbb.100918] [PMID: 21512225]

[29]   Weng Y, Xiang L, Matsuura A, Zhang Y, Huang Q, Qi J. Ganodermasides A and B, two novel anti-aging ergosterols from spores of a medicinal mushroom *Ganoderma lucidum* on yeast *via* UTH1 gene. Bioorg Med Chem 2010; 18(3): 999-1002.
[http://dx.doi.org/10.1016/j.bmc.2009.12.070] [PMID: 20093034]

[30]   Sabaratnam V, Kah-Hui W, Naidu M, Rosie David P. Neuronal health - can culinary and medicinal mushrooms help? J Tradit Complement Med 2013; 3(1): 62-8.
[http://dx.doi.org/10.4103/2225-4110.106549] [PMID: 24716157]

[31]   Seow SLS, Eik LF, Naidu M, David P, Wong KH, Sabaratnam V. *Lignosus rhinocerotis* (Cooke) Ryvarden mimics the neuritogenic activity of nerve growth factor *via* MEK/ERK1/2 signaling pathway in PC-12 cells. Sci Rep 2015; 5: 16349.
[http://dx.doi.org/10.1038/srep16349] [PMID: 26542212]

[32]   Ohta T, Kita T, Kobayashi N, *et al.* Scabronine A, a novel diterpenoid having potent inductive activity of the nerve growth factor synthesis, isolated from the mushroom, *Sarcodonscabrosus.* Tetrahedron Lett 1998; 39(34): 6229-32.
[http://dx.doi.org/10.1016/S0040-4039(98)01282-9]

[33]   Waters SP, Tian Y, Li YM, Danishefsky SJ. Total synthesis of (-)-scabronine G, an inducer of neurotrophic factor production. J Am Chem Soc 2005; 127(39): 13514-5.
[http://dx.doi.org/10.1021/ja055220x] [PMID: 16190712]

[34]   Lee IK, Yun BS, Kim JP, Ryoo IJ, Kim YH, Yoo ID. Neuroprotective activity of p-terphenyl leucomentins from the mushroom *Paxillus panuoides.* Biosci Biotechnol Biochem 2003; 67(8): 1813-6.
[http://dx.doi.org/10.1271/bbb.67.1813] [PMID: 12951520]

[35]   Sabaratnam V, Phan CW. Neuroactive components of culinary and medicinal mushrooms with potential to mitigate age-related neurodegenerative diseases.Discovery and Development of

Neuroprotective Agents from Natural Products. Elsevier 2018; pp. 401-13.
[http://dx.doi.org/10.1016/B978-0-12-809593-5.00010-0]

[36] Allen SJ, Watson JJ, Shoemark DK, Barua NU, Patel NK. GDNF, NGF and BDNF as therapeutic options for neurodegeneration. Pharmacol Ther 2013; 138(2): 155-75.
[http://dx.doi.org/10.1016/j.pharmthera.2013.01.004] [PMID: 23348013]

[37] Lee LY, Li IC, Chen WP, Tsai YT, Chen CC, Tung KC. Thirteen-Week Oral Toxicity Evaluation of Erinacine AEnriched Lion's Mane Medicinal Mushroom, *Hericium erinaceus* (Agaricomycetes), Mycelia in Sprague-Dawley Rats. Int J Med Mushrooms 2019; 21(4): 401-11.
[http://dx.doi.org/10.1615/IntJMedMushrooms.2019030320] [PMID: 31002635]

[38] Phan CW, David P, Tan YS, *et al*. Intrastrain comparison of the chemical composition and antioxidant activity of an edible mushroom, *Pleurotusgiganteus*, and its potent neuritogenic properties. ScientificWorldJournal 2014; 2014.

[39] Zhang CC, Cao CY, Kubo M, *et al*. Chemical Constituents from Hericium erinaceus Promote Neuronal Survival and Potentiate Neurite Outgrowth *via* the TrkA/Erk1/2 Pathway. Int J Mol Sci 2017; 18(8): 1659.
[http://dx.doi.org/10.3390/ijms18081659] [PMID: 28758954]

[40] Jang KJ, Kim HK, Han MH, *et al*. Anti-inflammatory effects of saponins derived from the roots of Platycodon grandiflorus in lipopolysaccharide☐stimulated BV2 microglial cells. Int J Mol Med. 2013 Jun;31(6):1357-66.
[http://dx.doi.org/10.3892/ijmm.2013.1330] [PMID: 23563392]

[41] Thu ZM, Myo KK, Aung HT, Clericuzio M, Armijos C, Vidari G. Bioactive Phytochemical Constituents of Wild Edible Mushrooms from Southeast Asia. Molecules 2020; 25(8): 1972.
[http://dx.doi.org/10.3390/molecules25081972] [PMID: 32340227]

[42] Hong Y, Ying T. Characterization of a chitin-glucan complex from the fruiting body of Termitomyces albuminosus (Berk.) Heim. Int J Biol Macromol 2019; 134: 131-8.
[http://dx.doi.org/10.1016/j.ijbiomac.2019.04.198] [PMID: 31063786]

[43] Choi JH, Maeda K, Nagai K, *et al*. Termitomycamides A to E, fatty acid amides isolated from the mushroom *Termitomyces titanicus*, suppress endoplasmic reticulum stress. Org Lett 2010; 12(21): 5012-5.
[http://dx.doi.org/10.1021/ol102186p] [PMID: 20936815]

[44] Choi YJ, Yang HS, Jo JH, *et al*. Anti-amnesic effect of fermented *Ganoderma lucidum* water extracts by lactic acid bacteria on scopolamine-induced memory impairment in rats. Prev Nutr Food Sci 2015; 20(2): 126-32.
[http://dx.doi.org/10.3746/pnf.2015.20.2.126] [PMID: 26176000]

[45] Antonelli M, Donelli D, Barbieri G, Valussi M, Maggini V, Firenzuoli F. Forest Volatile Organic Compounds and Their Effects on Human Health: A State-of-the-Art Review. Int J Environ Res Public Health 2020; 17(18): 6506.
[http://dx.doi.org/10.3390/ijerph17186506] [PMID: 32906736]

[46] Kawagishi H, Zhuang C, Yunoki R. Compounds for dementia from *Hericiumerinaceum*. Drugs Future 2008; 33(2): 149.
[http://dx.doi.org/10.1358/dof.2008.033.02.1173290]

[47] Lizarme Y, Wangsahardja J, Marcolin GM, Morris JC, Jones NM, Hunter L. Synthesis and neuroprotective activity of dictyoquinazol A and analogues. Bioorg Med Chem 2016; 24(7): 1480-7.
[http://dx.doi.org/10.1016/j.bmc.2016.02.016] [PMID: 26906473]

[48] Brondz I, Ekeberg D, Høiland K, Bell DS, Annino AR. The real nature of the indole alkaloids in *Cortinarius infractus*: evaluation of artifact formation through solvent extraction method development. J Chromatogr A 2007; 1148(1): 1-7.
[http://dx.doi.org/10.1016/j.chroma.2007.02.074] [PMID: 17391682]

[49]   Nakamichi N, Taguchi T, Hosotani H, *et al.* Functional expression of carnitine/organic cation transporter OCTN1 in mouse brain neurons: possible involvement in neuronal differentiation. Neurochem Int 2012; 61(7): 1121-32.
[http://dx.doi.org/10.1016/j.neuint.2012.08.004] [PMID: 22944603]

[50]   Zhang R, Xu S, Cai Y, Zhou M, Zuo X, Chan P. Ganoderma lucidum Protects Dopaminergic Neuron Degeneration through Inhibition of Microglial Activation. Evid Based Complement Alternat Med. 2011;2011:156810.
[http://dx.doi.org/10.1093/ecam/nep075] [PMID: 19617199]

[51]   Ling-Sing Seow S, Naidu M, David P, Wong KH, Sabaratnam V. Potentiation of neuritogenic activity of medicinal mushrooms in rat pheochromocytoma cells. BMC Complement Altern Med 2013; 13(1): 157.
[http://dx.doi.org/10.1186/1472-6882-13-157] [PMID: 23822837]

[52]   Shi XW, Liu L, Gao JM, Zhang AL. Cyathane diterpenes from Chinese mushroom *Sarcodon scabrosus* and their neurite outgrowth-promoting activity. Eur J Med Chem 2011; 46(7): 3112-7.
[http://dx.doi.org/10.1016/j.ejmech.2011.04.006] [PMID: 21530015]

[53]   Trovato A, Siracusa R, Di Paola R, *et al.* Redox modulation of cellular stress response and lipoxin A4 expression by *Coriolus versicolor* in rat brain: Relevance to Alzheimer's disease pathogenesis. Neurotoxicology 2016; 53: 350-8.
[http://dx.doi.org/10.1016/j.neuro.2015.09.012] [PMID: 26433056]

[54]   Nishina A, Kimura H, Sekiguchi A, Fukumoto RH, Nakajima S, Furukawa S. Lysophosphatidylethanolamine in *Grifola frondosa* as a neurotrophic activator *via* activation of MAPK. J Lipid Res 2006; 47(7): 1434-43.
[http://dx.doi.org/10.1194/jlr.M600045-JLR200] [PMID: 16614393]

[55]   Hage S, Kienlen-Campard P, Octave JN, Quetin-Leclercq J. *In vitro* screening on β-amyloid peptide production of plants used in traditional medicine for cognitive disorders. J Ethnopharmacol 2010; 131(3): 585-91.
[http://dx.doi.org/10.1016/j.jep.2010.07.044] [PMID: 20673795]

[56]   Solanki I, Parihar P, Mansuri ML, Parihar MS. Flavonoid-based therapies in the early management of neurodegenerative diseases. Adv Nutr 2015; 6(1): 64-72.
[http://dx.doi.org/10.3945/an.114.007500] [PMID: 25593144]

[57]   Apostolova LG. Alzheimer Disease. Continuum (Minneap Minn). 2016 Apr;22(2 Dementia):419-34.
[http://dx.doi.org/10.1212/CON.0000000000000307] [PMID: 27042902]

[58]   Demetrius LA, Driver J. Alzheimer's as a metabolic disease. Biogerontology 2013; 14(6): 641-9.
[http://dx.doi.org/10.1007/s10522-013-9479-7] [PMID: 24249045]

[59]   Ahuja M, Patel M, Majrashi M, Mulabagal V, Dhanasekaran M. Centellaasiatica, an Ayurvedic Medicinal Plant, Prevents the Major Neurodegenerative and Neurotoxic Mechanisms Associated with Cognitive Impairment.Medicinal Plants and Fungi: Recent Advances in Research and Development. Singapore: Springer 2017; pp. 3-48.
[http://dx.doi.org/10.1007/978-981-10-5978-0_1]

[60]   Phan CW, David P, Sabaratnam V. Edible and medicinal mushrooms: emerging brain food for the mitigation of neurodegenerative diseases. J Med Food 2017; 20(1): 1-10.
[http://dx.doi.org/10.1089/jmf.2016.3740] [PMID: 28098514]

[61]   Ćilerdžić J, Galić M, Vukojević J, Stajic M. *Pleurotusostreatus* and *Laetiporussulphureus* (Agaricomycetes): Possible agents against Alzheimer and Parkinson diseases. Int J Med Mushrooms 2019; 21(3): 275-89.
[http://dx.doi.org/10.1615/IntJMedMushrooms.2019030136] [PMID: 31002611]

[62]   Wong KH, Sabaratnam V, Naidu M, Keynes R. Activity of aqueous extracts of lion's mane mushroom Hericiumerinaceus (Bull.: Fr.) Pers.(Aphyllophoromycetideae) on the neural cell line NG108-15.

International Journal of Medicinal Mushrooms, 2007, 9(1).
[http://dx.doi.org/10.1615/IntJMedMushr.v9.i1.70]

[63]   Lai PL, Naidu M, Sabaratnam V, *et al.* Neurotrophic properties of the Lion's mane medicinal mushroom, *Hericium erinaceus* (Higher Basidiomycetes) from Malaysia. Int J Med Mushrooms 2013; 15(6): 539-54.
[http://dx.doi.org/10.1615/IntJMedMushr.v15.i6.30] [PMID: 24266378]

[64]   Mori K, Obara Y, Hirota M, *et al.* Nerve growth factor-inducing activity of *Hericium erinaceus* in 1321N1 human astrocytoma cells. Biol Pharm Bull 2008; 31(9): 1727-32.
[http://dx.doi.org/10.1248/bpb.31.1727] [PMID: 18758067]

[65]   Mori K, Inatomi S, Ouchi K, Azumi Y, Tuchida T. Improving effects of the mushroom Yamabushitake (*Hericiumerinaceus*) on mild cognitive impairment: A double□blind placebo□controlled clinical trial. Phytotherapy Research. An International Journal Devoted to Pharmacological and Toxicological Evaluation of Natural Product Derivatives 2009; 23(3): 367-72.
[PMID: 18844328]

[66]   Mori K, Obara Y, Moriya T, Inatomi S, Nakahata N. Effects of *Hericium erinaceus* on amyloid β(25-35) peptide-induced learning and memory deficits in mice. Biomed Res 2011; 32(1): 67-72.
[http://dx.doi.org/10.2220/biomedres.32.67] [PMID: 21383512]

[67]   Kuo HC, Lu CC, Shen CH, *et al. Hericium erinaceus* mycelium and its isolated erinacine A protection from MPTP-induced neurotoxicity through the ER stress, triggering an apoptosis cascade. J Transl Med 2016; 14(1): 78.
[http://dx.doi.org/10.1186/s12967-016-0831-y] [PMID: 26988860]

[68]   Karami A, Eyjolfsdottir H, Vijayaraghavan S, *et al.* Changes in CSF cholinergic biomarkers in response to cell therapy with NGF in patients with Alzheimer's disease. Alzheimers Dement 2015; 11(11): 1316-28.
[http://dx.doi.org/10.1016/j.jalz.2014.11.008] [PMID: 25676388]

[69]   Elufioye TO, Berida TI, Habtemariam S. Plants-Derived Neuroprotective Agents: Cutting the Cycle of Cell Death through Multiple Mechanisms. Evid Based Complement Alternat Med 2017; 2017: 3574012.
[http://dx.doi.org/10.1155/2017/3574012] [PMID: 28904554]

[70]   Nagano M, Shimizu K, Kondo R, *et al.* Reduction of depression and anxiety by 4 weeks *Hericium erinaceus* intake. Biomed Res 2010; 31(4): 231-7.
[http://dx.doi.org/10.2220/biomedres.31.231] [PMID: 20834180]

[71]   Zhang J, An S, Hu W, *et al.* The Neuroprotective Properties of Hericium erinaceus in Glutamate-Damaged Differentiated PC12 Cells and an Alzheimer's Disease Mouse Model. Int J Mol Sci 2016; 17(11): 1810.
[http://dx.doi.org/10.3390/ijms17111810] [PMID: 27809277]

[72]   Li Z, Chen X, Lu W, *et al.* Anti-oxidative stress activity is essential for *Amanita caesarea* mediated neuroprotection on glutamate-induced apoptotic HT22 cells and an Alzheimer's disease mouse model. Int J Mol Sci 2017; 18(8): 1623.
[http://dx.doi.org/10.3390/ijms18081623] [PMID: 28749416]

[73]   Jeong Y, Jung M, Kim MJ, Hwang CH. A 4-nitroquinolineoxide-induced *Pleurotuseryngii* mutant variety increases Pin1 expression in rat brain. J Med Food 2017; 20(1): 65-70.
[http://dx.doi.org/10.1089/jmf.2016.3809] [PMID: 28098518]

[74]   Fernández-Sanz P, Ruiz-Gabarre D, García-Escudero V. Modulating effect of diet on Alzheimer's disease. Diseases 2019; 7(1): 12.
[http://dx.doi.org/10.3390/diseases7010012] [PMID: 30691140]

[75]   Scarmeas N, Stern Y, Tang MX, Mayeux R, Luchsinger JA. Mediterranean diet and risk for Alzheimer's disease. Ann Neurol 2006; 59(6): 912-21.
[http://dx.doi.org/10.1002/ana.20854] [PMID: 16622828]

[76]　Chong PS, Fung ML, Wong KH, Lim LW. Therapeutic potential of *Hericiumerinaceus* for depressive disorder. Int J Mol Sci 2020; 21(1): 163.
[http://dx.doi.org/10.3390/ijms21010163]

[77]　Bennett L, Kersaitis C, Macaulay SL, *et al.* Vitamin D2-enriched button mushroom (*Agaricus bisporus*) improves memory in both wild type and APPswe/PS1dE9 transgenic mice. PLoS One 2013; 8(10): e76362.
[http://dx.doi.org/10.1371/journal.pone.0076362] [PMID: 24204618]

[78]　Phan CW, David P, Naidu M, Wong KH, Sabaratnam V. Therapeutic potential of culinary-medicinal mushrooms for the management of neurodegenerative diseases: diversity, metabolite, and mechanism. Crit Rev Biotechnol 2015; 35(3): 355-68.
[http://dx.doi.org/10.3109/07388551.2014.887649] [PMID: 24654802]

[79]　Phan CW, Wong WL, David P, Naidu M, Sabaratnam V. *Pleurotus giganteus* (Berk.) Karunarathna & K.D. Hyde: Nutritional value and in *vitro* neurite outgrowth activity in rat pheochromocytoma cells. BMC Complement Altern Med 2012; 12(1): 102.
[http://dx.doi.org/10.1186/1472-6882-12-102] [PMID: 22812497]

[80]　Paidi RK, Nthenge-Ngumbau DN, Singh R, Kankanala T, Mehta H, Mohanakumar KP. Mitochondrial deficits accompany cognitive decline following single bilateral intracerebroventricular streptozotocin. Curr Alzheimer Res 2015; 12(8): 785-95.
[http://dx.doi.org/10.2174/1567205012666150710112618] [PMID: 26159195]

[81]　Huang S, Mao J, Ding K, *et al.* Polysaccharides from *Ganoderma lucidum* promote cognitive function and neural progenitor proliferation in mouse model of Alzheimer's disease. Stem Cell Reports 2017; 8(1): 84-94.
[http://dx.doi.org/10.1016/j.stemcr.2016.12.007] [PMID: 28076758]

[82]　Bakir T, Karadeniz M, Unal S. Investigation of antioxidant activities of *Pleurotus ostreatus* stored at different temperatures. Food Sci Nutr 2018; 6(4): 1040-4.
[http://dx.doi.org/10.1002/fsn3.644] [PMID: 29983968]

[83]　Lee SC, Im NK, Jeong HY, Choi EH, Jeon SM, Jeong GS. Neuroprotective effects of ethanol extract of *Ganoderma lucidum* L. on murine hippocampal cells. Korean J Pharmacogn 2014; 45(2): 161-7.

[84]　Zhong L, Yan P, Lam WC, Yao L, Bian Z. *Coriolus Versicolor* and *Ganoderma Lucidum* Related Natural Products as an Adjunct Therapy for Cancers: A Systematic Review and Meta-Analysis of Randomized Controlled Trials. Front Pharmacol 2019; 10: 703.
[http://dx.doi.org/10.3389/fphar.2019.00703] [PMID: 31333449]

[85]　Scuto M, Di Mauro P, Ontario ML, *et al.* Nutritional Mushroom Treatment in Meniere's Disease with *Coriolus versicolor*: A Rationale for Therapeutic Intervention in Neuroinflammation and Antineurodegeneration. Int J Mol Sci 2019; 21(1): 284.
[http://dx.doi.org/10.3390/ijms21010284] [PMID: 31906226]

[86]　N. S., &Sirajudeen, K. N. S. (2020). Natural Products and Their Bioactive Compounds: Neuroprotective Potentials against Neurodegenerative Diseases. Evid Based Complement Alternat Med 2020.

[87]　Chandra G, Shenoi RA, Anand R, Rajamma U, Mohanakumar KP. Reinforcing mitochondrial functions in aging brain: An insight into Parkinson's disease therapeutics. J Chem Neuroanat 2019; 95: 29-42.
[http://dx.doi.org/10.1016/j.jchemneu.2017.12.004] [PMID: 29269015]

[88]　Fahn S. Description of Parkinson's disease as a clinical syndrome. Ann N Y Acad Sci 2003; 991: 1-14.
[http://dx.doi.org/10.1111/j.1749-6632.2003.tb07458.x] [PMID: 12846969]

[89]　Singh A, Verma P, Balaji G, Samantaray S, Mohanakumar KP. Nimodipine, an L-type calcium channel blocker attenuates mitochondrial dysfunctions to protect against 1-methyl-4-phenyl-1,2-3,6-tetrahydropyridine-induced Parkinsonism in mice. Neurochem Int 2016; 99: 221-32.

[http://dx.doi.org/10.1016/j.neuint.2016.07.003] [PMID: 27395789]

[90]   Singh A, Verma P, Raju A, Mohanakumar KP. Nimodipine attenuates the parkinsonian neurotoxin, MPTP-induced changes in the calcium binding proteins, calpain and calbindin. J Chem Neuroanat 2019; 95: 89-94.
[http://dx.doi.org/10.1016/j.jchemneu.2018.02.001] [PMID: 29427747]

[91]   Zhu WW, Liu ZL, Xu HW, *et al.* Effect of the oil from ganoderma lucidum spores on pathological changes in the substantia nigra and behaviors of MPTP-treated mice. Di Yi Jun Yi Da Xue Xue Bao. 2005 Jun;25(6):667-71.
[PMID: 15958304]

[92]   Venkatesh Gobi V, Rajasankar S, Ramkumar M, *et al. Agaricus blazei* extract attenuates rotenone-induced apoptosis through its mitochondrial protective and antioxidant properties in SH-SY5Y neuroblastoma cells. Nutr Neurosci 2018; 21(2): 97-107.
[http://dx.doi.org/10.1080/1028415X.2016.1222332] [PMID: 27646574]

[93]   Venkatesh Gobi V, Rajasankar S, Ramkumar M, *et al. Agaricus blazei* extract abrogates rotenone-induced dopamine depletion and motor deficits by its anti-oxidative and anti-inflammatory properties in Parkinsonic mice. Nutr Neurosci 2018; 21(9): 657-66.
[http://dx.doi.org/10.1080/1028415X.2017.1337290] [PMID: 28628424]

CHAPTER 7

Advances in Experimental Animal Models Provide Insights into Different Etiology and Mechanism of Multiple Sclerosis to Design Therapeutics

Sourodip Sengupta and **Jayasri Das Sarma**[*]

Department of Biological Sciences, Indian Institute of Science Education and Research Kolkata (IISER-K), Mohanpur, Nadia, West Bengal-741246, India

Abstract: Myelin covering of axons in the central and peripheral nervous system helps in faster propagation of neuronal action potentials. Demyelination is a neurodegenerative process in which the axons lose their myelin coverings, exposing the axons to surroundings and leading to a reduction in neuron-to-neuron communication. Several demyelinating diseases exist in humans, and one of the most frequently occurring demyelinating disease of the CNS is multiple sclerosis (MS). Although more than 2.3 million people suffer from MS globally, the disease etiology is still unknown, impeding the development of effective therapeutics. The available treatments are based on disease-modifying therapy to reduce or moderate the symptoms and slow the disease progression; however, none can cure the disease. One key to better design therapeutics is to understand the cellular and molecular mechanisms of MS by developing reliable model systems. Human studies have their own limitations, such as limited access to patient tissues. Moreover, genetic variability makes it difficult to identify the triggers of MS. This calls for the development of reliable experimental animal models to understand MS pathogenesis better. There is no exclusive experimental model that covers the entire gamut of the disease. In this chapter, we will discuss experiment autoimmune encephalomyelitis (EAE), Theiler's murine encephalomyelitis virus (TMEV), and mouse hepatitis virus (MHV)-induced models of demyelination that mimic specific histopathological and neurobiological aspects of multiple sclerosis. The present understanding of MS as an autoimmune disease mediated by self-reactive T-cells comes mainly from studies on the EAE model. Further, viral-induced demyelination models have provided valuable insights into a better understanding of MS. Studies in the TMEV model have demonstrated molecular mimicry and epitope spreading as major mechanisms of virus-induced neuroinflammation. Our knowledge of immune-mediated CNS damage has been further enhanced by studies on MHV-induced neuroinflammatory demyelination, suggesting macrophage-mediated myelin stripping in neurodegeneration. While the limitations of these models of MS are obvious, appropriate use of this model has led to the development of clinically useful drugs for the treatment of this devastating disease.

[*] **Corresponding author Jayasri Das Sarma:** Indian Institute of Science Education and Research Kolkata IISER-K), Department of Biological Sciences, Mohanpur, Nadia, West Bengal-741246, India; Tel: +91 7003514069; E-mail: dassarmaj@iiserkol.ac.in

Sachchida Nand Rai (Ed.)

Keywords: Demyelination, EAE, MHV, Multiple sclerosis, Neurodegeneration, Neuroinflammation, TMEV.

INTRODUCTION

The brain, which forms part of the Central nervous system (CNS), is an important organ that controls other organs through nerve connectivity and conduction of nerve impulses [1].

Myelin covering of axons in the central and peripheral nervous system (PNS) helps in faster propagation of neuronal action potentials. Demyelination is a neurodegenerative process in which the axons lose their myelin coverings, exposing the axons to surroundings and leading to a reduction in neuron-t- -neuron communication [2]. Several demyelinating diseases exist in humans, including:

- Optic neuritis, Neuromyelitis Optica (Devic's disease)—inflammation of the optic nerve.

- Transverse myelitis—inflammation of the spinal cord.

- Acute disseminated encephalomyelitis—inflammation of the brain and spinal cord.

One of the relatively common demyelinating diseases of the CNS is multiple sclerosis (MS). MS is a chronic demyelinating disease of the CNS where the myelin sheath gets damaged and forms scar tissue termed sclerosis. The resulting nerve damage disrupts the transmission of nerve impulses [3].

Earlier notions visualized the brain as an immunologically protected site with its immune surveillance system in the form of glial cells and restricted communications with the peripheral immune system(PIS). However, the previous notion has been proved wrong with the understanding that there exists a bi-directional communication between the PIS and the brain [1]. In general, inflammation in the CNS, commonly termed neuroinflammation, serves as a protective mechanism [1]. However, prolonged neuroinflammation is a responsible factor for the onset of neurodegeneration and neuronal losses associated with it. Neurodegeneration is a condition in which neurons suffer structural and functional alterations, resulting in reduced survival and increased neuronal death [1].

Recent findings showed the presence of inflammation in the brain regions of autopsy samples from MS patients, indicating a link between inflammation and neurodegeneration [4]. Further studies revealed the presence of antibody deposits within MS lesions in the brain from the progressive disease phase [5]. These observations are indicative of a connection between neuroinflammation and neurodegeneration in MS. However, it is unclear whether the primary cause for neurodegeneration is neuroinflammation or it is only a secondary effect.

CLASSIFICATION OF MS

Globally, more than 2.3 million people suffer from MS. Most diagnosed patients fall within the age bracket of 20-50 years, and about two-thirds are women. Periods of active MS symptoms are called relapse or attacks, while the remission period follows no attack. In relapse-remitting MS (RR-MS), patients have attacks followed by partial or total remission that may extend between months to years. Most people are diagnosed at the relapse-remitting stage [3].

In primary progressive (PP-MS) disease conditions, patients miss the remission phase, and symptoms worsen steadily from the onset, which is usually observed in 10-15% of cases [3]. PP-MS is generally seen in older patients, suggesting that age could be a determining factor for disease progression [6]. In the majority of RR-MS patients, the disease may become progressive after an initial remission period, as seen in secondary progressive (SP) MS [2, 7].

ETIOLOGY OF MS

The exact cause of MS is not known yet, and a multitude of factors such as genetic, environmental, and autoimmunity can be responsible for the disease onset. The rate of MS incidence is relatively low in childhood. Chances of MS increases post 18 years of age, with maximum occurrence observed in people falling between 25-35 years of age (women are at a higher risk than men). The appearance of MS slowly declines with age, becoming rare at age 50 and older [8].

While MS is prevalent in the US, Canada, Europe, New Zealand, and parts of Australia, it is relatively scarce in Asia and is generally limited to the tropics and subtropics. In temperate regions, the incidence of MS increases with latitude [8]. Epidemiological studies suggest a correlation between disease prevalence and exposure to sunlight [9]. Sunlight exposure is a significant source of vitamin D for most people. As the amount of sunlight decreases at higher latitudes, a greater incidence of MS could be due to vitamin D deficiency. A fascinating observation is that the risk of MS declines among individuals migrating from high- to low-risk areas but does not necessarily increase with migration in the opposite direction

[9]. Overall, these studies implicate that environmental factors might play an essential role in determining MS risk but are yet to be fully understood.

Genetic factors could also determine susceptibility to MS. It has been observed that about 10% of patients suffering from MS have relatives diagnosed with MS [10]. Also, the higher frequency of certain major histocompatibility complex (MHC) or human leukocyte antigen (HLA) regions in MS patients suggests that susceptibility to MS is MHC dependent. In particular, certain HLA-II alleles are found to be associated with MS susceptibility. One such candidate is the HLA-DRB1*15:01 haplotypes. Around 30% of MS patients from the US and northern parts of Europe are found to express this gene variant [10]. Further, the DRB1 haplotype is known to be regulated epigenetically. Hypomethylation is associated with increased expression of this haplotype in monocytes of DRB1*15:01 carriers [11].

Several viruses such as the Epstein-Barr virus [12] and human herpesviruses [13] have been found in the CNS of MS patients; however, it is uncertain whether these viruses are endogenic or pathogens prompting the disease. Furthermore, the presence of an increased amount of IgG antibody manifest as oligoclonal bands (OGBs) from the brain and cerebrospinal fluid (CSF) of more than 90% of MS patients is suggestive that MS might be infectious [14]. Usually, increased amounts of OGBs and IgG are associated with infectious diseases of CNS. Moreover, the IgG molecules are highly specific against the agent that cause the disease [15]. These findings indicate that the IgGs in the MS brain are antibodies directed against some infectious agents.

MS is also considered as an autoimmunity disorder because genetically predisposed people, when exposed to environmental factors, suffer from T-cell autoreactivity to self-antigens [16]. Additionally, the contribution of innate immune cells in mediating MS pathogenesis has also gained attention as microglial cells have the potential to initiate anti-inflammatory or pro-inflammatory response upon interaction with myelin [17].

CLINICAL SYMPTOMS AND DIAGNOSIS

Demyelination and axonal loss are the two major hallmarks of MS responsible for permanent neurological damage.MS diagnosis is not always easy as the symptoms observed are often confused with other CNS diseases. Moreover, symptoms may come and go over time in a person delaying the diagnosis. The presence of multiple scar tissue patches in the CNS detectable in an MRI scan, along with at least two pieces of evidence of an attack, can confirm MS in a patient [3].

Symptoms vary significantly in a patient and may be mild, moderate, or severe. Patients may suffer from numbness in hands and feet, muscle weakness, poor balance, and coordination. Severe symptoms include slurred speech, temporary or permanent paralysis, and even loss of vision. Mood changes, such as depression and memory loss, are also reported in MS [3].

PRESENT THERAPEUTICS AND LIMITATIONS

MS is a complex disease with a multitude of responsible factors and heterogeneity in disease symptoms. The available treatment is based on disease-modifying therapy to reduce or moderate the symptoms and slow the disease progression. A list of all FDA approved drugs with details of their mode of action, dosage, and side-effects are available in a consensus paper released by the MS Coalition [18]. All of these medications come along with their warnings and precautions, as they can have severe side effects. However, none can cure the disease but only helps to improve the quality of life of the patients. Other management strategies involve lifestyle changes that include physical exercises, speech, and cognitive therapy, counseling, and diet plan [3]. One key to better design therapeutics is to understand the cellular and molecular mechanisms of MS by developing reliable model systems.

EXPERIMENTAL ANIMAL MODELS TO UNDERSTAND THE MECHANISMS OF MS

There are several limitations of conducting studies with human samples, such as limited access to patient tissues and unavailability of biopsy samples, and autopsy samples are usually from patients with chronic disease phase. Moreover, disease heterogeneity due to genetic complexity, variability in the pathology, and symptoms make it difficult to identify triggers of MS. Therefore, appropriate experimental animal models are needed to understand the pathology of MS and overcome therapeutic challenges [2].

There is no one experimental model that covers all the clinical and pathological aspects of the disease. Several models are available that allow studying individual aspects and mechanisms relevant to the disease pathogenesis.

EXPERIMENTAL AUTOIMMUNE ENCEPHALOMYELITIS (EAE)

Experimental Autoimmune Encephalomyelitis, or EAE as a model to study inflammation in MS was first described in monkeys in 1933 and remained a popular and widely used model [19, 20]. EAE has been induced in a wide range of rodent and primate models that reproduce specific aspects of the human disease [21].

In the classical EAE model, animals are immunized with subcutaneous injection of a neuro antigen with an adjuvant. In mice, EAE can be induced with MBP (myelin basic protein)–PLP fusion protein MP4, proteolipid protein (PLP) peptide 178–191, or myelin oligodendrocyte glycoprotein (MOG) peptide 35–55 [22].

Induction of EAE with MP4 results in dynamic and time-dependent infiltration of inflammatory cells characterized by macrophages, dendritic cells, CD[4+], and CD[8+] T cells, along with B cells. In MOG peptide and PLP models, macrophages, along with CD4 expressing T lymphocytes, are the predominant cellular infiltrates [23]. The three models also show region-specificity of spinal cord degeneration along with differential demyelination: acute, transient, and chronic [22]. Each model produces distinct CNS inflammation patterns and spinal cord pathology, thereby serves as a useful tool in reproducing and studying the structural–morphological diversity of MS.

The present understanding of MS as an autoimmune disease mediated by self-reactive T-cells comes essentially from studies on the EAE model. Also, by using EAE models, we have been able to delineate the important role played by the humoral immune system in demyelination [21].

It has long been believed that Th1 cells, a subset of CD[4+] T cells, have a pathogenic role in EAE. Indeed, Th1 cells expressing IFN-α/β can induce EAE and have been found in increased amounts in the blood and CSF of MS patients [24]. On the other hand, mice deficient in IFN-α/β, the main effectors' cytokine of Th1 cells, could still develop EAE suggesting that Th1 is not crucial to generate EAE [24]. The importance of IL-17 secreting Th17 cells have been shown in adoptive transfer studies, wherein MOG-specific Th17 cells induced robust EAE [25]. These findings highlight the complex role played by different subsets of CD[4+] cells in demyelination.

Experimental autoimmune encephalomyelitis serves as an essential instrument to study immune-mediated neurodegeneration and for assessing the potential of therapeutic approaches in disease prevention. Past studies in EAE models have led to the discovery of drugs such as glatiramer acetate (GA) and natalizumab [21].

GA is a copolymer of four amino acid residues that was originally developed to mimic MBP for the induction of EAE. Surprisingly, the administration of GA failed to induce EAE and prevented disease progression when given with MBP

[24]. GA is the only available antigen therapy that seems to block myelin-damaging T-cells through a mechanism not entirely understood [18]. GA is effective in treating RR-MS and probably acts by interfering with antigen presentation and promoting self-tolerance [24].

Natalizumab is a monoclonal antibody against a4b1 integrins found on lymphocytes and restricts their trafficking into the CNS, thereby preventing EAE progression [24]. It is used in the treatment of relapsing MS [18].

A significant difference between EAE and MS is that while MS is a spontaneous disease, EAE is induced by active sensitization with brain tissue antigens in the presence of potent immune adjuvants to produce disease [21]. It is doubtful that such strong immune reactions occur under physiological conditions. Also, EAE is studied mainly in genetically homogeneous populations of animals [21]. Thus, a single EAE model fails to cover the heterogeneity in the genetic make-up of MS patients. Moreover, it has not been possible to generate an EAE model for primary progressive MS [21]. Studies in EAE models have underestimated the relevance of cytotoxic CD^{8+} T cells in neuroinflammation. While the EAE model has its limitations, the appropriate use of this model will continue to improve our understanding of this devastating disease.

EXPERIMENTAL VIRAL-INDUCED MODEL OF MS

In the virus-induced model of demyelination, infection of CNS is a precondition for the onset of the disease. Also, the cause-effect association makes this a suitable tool for dissecting the underlying mechanism of demyelinating diseases [26].

In humans, viruses can cause chronic demyelinating diseases. Progressive multifocal leukoencephalopathy (PML) and subacute sclerosing panencephalitis (SSPE) are two such diseases [27]. PML and SSPE are caused by an infection with papovavirus (DNA virus) and measles virus (RNA virus), respectively. Such observations have fuelled interest in understanding the mechanisms of viral-induced demyelination.

TMEV MODEL

Theiler's murine encephalomyelitis virus (TMEV) was first described by Theiler in 1934. TMEV belongs to the family Picornaviridae having a single-stranded positive-sense RNA genome and causes neurological and enteric diseases in susceptible strains of mice [28, 29].

Depending upon neurovirulence, TMEV strains are divided into two subgroups, GDVII and TO. The GDVII subgroup is neurovirulent. The TO subgroup is less virulent, and infection of susceptible mouse strains (SJL/J) with TO strains promotes a biphasic disease consisting of acute and chronic inflammation. This chronic period marked by the presence of demyelination serves as a model for primary progressive MS [10].

In the TMEV model, the early acute disease phase is associated with replication of the virus in the grey matter and extensive infiltration of CNS with mononucleated cells, comprising mainly of CD4 and CD8-expressing T cells and monocytes or macrophages. Throughout the acute period, inflammation is localized to the grey matter region, and the white matter remains untouched. Although viral titer considerably falls by 12 days post-infection, TMEV continues to persist in macrophages and glial cells (microglia, astrocytes, and oligodendrocytes) throughout the animal's lifespan [10].

The chronic demyelinating disease (usually occurs at 30-35 days post-infection) in mice infected with TMEV bears close similarities with MS in humans. Both exhibit inflammatory infiltrates consisting largely of T- and B-lymphocytes in addition to monocytes or macrophages. Also, demyelinating lesions with oligodendroglia damage and spinal cord degeneration indicative of nervous dysfunction are observed in both [10].

The importance of the axonal injury and spinal cord demyelination, in addition to the role of epitope spreading from viral antigen to self-myelin epitopes, are some of the discoveries of studies conducted with the TMEV model [30].

Chronic viral persistence leads to the release of inflammatory mediators by activated lymphocytes and monocytes/macrophages. Prolonged-release of cytotoxic factors such as cytokines, nitric oxide (NO) metabolites, and reactive oxygen species (ROS) causes damage to the myelin sheath. The myelin damage initiates the release of self-myelin antigens that are then engulfed by the CNS resident APCs (Antigen Presenting Cells) and offered to the T cells resulting in myelin-specific autoreactive CD^{4+} T cells. This phenomenon is also known as epitope spreading [30]. With the onset of epitope spreading, CD4 expressing T cells autoreactive against the myelin proteolipid protein (PLP) appears in the late chronic phase, and as the disease advances, CD^{4+} T cell-induced autoimmunity to several myelin epitopes occur [30].

Additionally, viruses may induce autoimmunity *via* molecular mimicry in which immunological cross-reactivity between viral epitope and self-antigenic epitopes occur. In TMEV infection, molecular mimicry was observed between viral peptides and myelin-associated lipids such as galactocerebroside (GALC).

Moreover, TMEV infection contributes to demyelination by inducing the production of antibodies reactive against myelin components [30]. Thus, epitope spreading and molecular mimicry serve as the underlying mechanisms for TMEV-induced demyelination.

The last few decades have witnessed immense interest in addressing the underlying mechanism of axonal injury. Currently, two models exist that explain the axonal damage [30]. In the inside-out model, axonal injury occurs before demyelination, while in the outside-in model, axonal loss follows primary demyelination induced by the infiltrating lymphocytes Fig. (**1**). However, the exact mechanism of axonal damage in TMEV infection is still unclear.

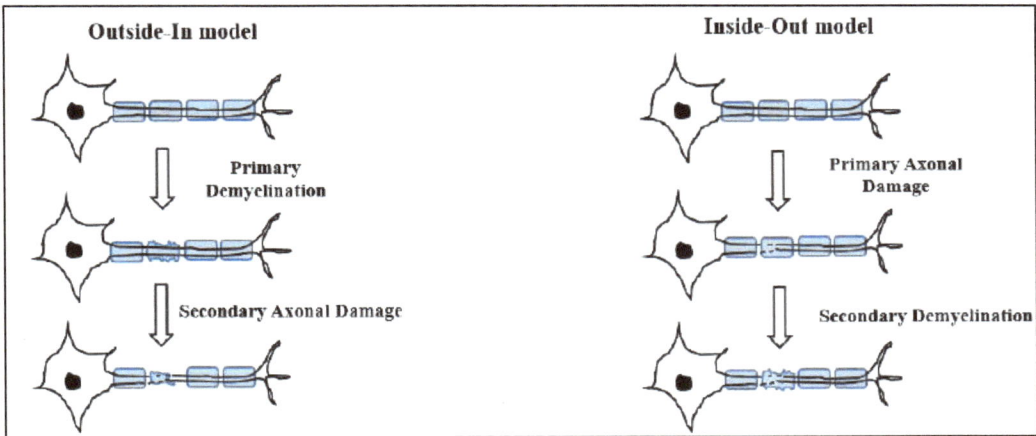

Fig. (1). Mechanism of axonal injury and demyelination in the TMEV model. The outside-in model proposes that myelin represents the primary target for injury. The infiltrating lymphocytes first attack the myelin sheath (outside), and the damage extends inwards to the axons. In the inside-out model, the initial injuries to the axon (inside) extend outward to the myelin covering.

The TMEV model has improved our understanding of the disease and aided in the development of interferon beta (IFN- α/β) as one of the most successful therapies for MS [18]. EAE model had first shown that type 1 IFN could reduce disease severity and was later approved for MS treatment. Although IFN- α/β treatment is effective in the majority of cases, around 30% of patients show no sign of improvement with IFN- α/β [24].

Analysis of IFN-α/β treatment in the TMEV model has improved our understanding of treatment with IFN-α/β. TMEV model has demonstrated the ability of IFN-α/β to modulate the expression of MHC class 1 molecule (provides stimulatory signals to CD^{8+} T cells) in the CNS [30]. Further, studies conducted in the TMEV model have shown that short-term treatment with IFN-α/β leads to remyelination in the spinal cord white matter. Paradoxically, the long-term treatment promotes demyelination, indicating that the effectiveness of type 1 IFN depends on treatment length [31]. Despite having limitations, the TMEV model remains critical to our understanding of the etiological variability associated with MS. Fig. (1) shows the CNS pathology of mice infected with a neurotropic strain of MHV.

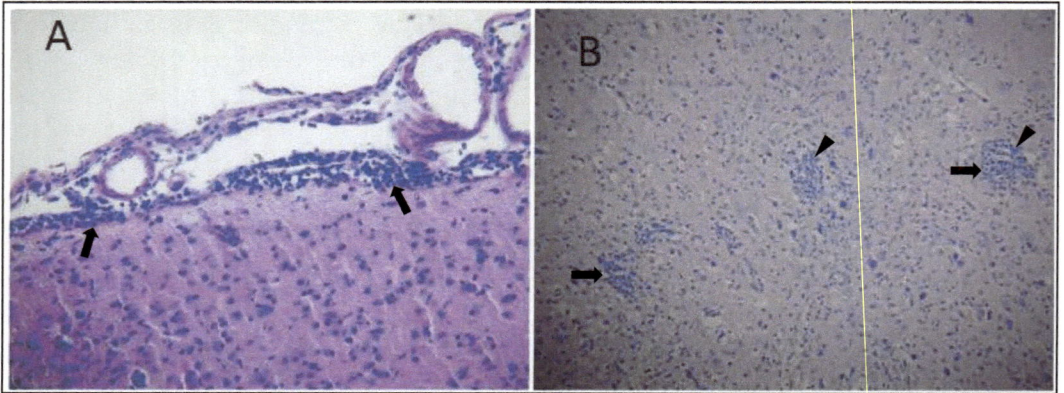

Fig. (2). CNS pathology of mice infected with a neurotropic strain of MHV. Presence of meningitis **(A)** and encephalitis **(B)** in 5 μm thick brain serial sections of mice at 7 days p.i upon staining with haematoxylin and eosin. Acute inflammation is associated with the accumulation of inflammatory cells in the meninges and brain parenchyma. Microglia or macrophages are the predominant inflammatory cells often seen as nodules (arrowheads in B) in the parenchyma. Arrows indicate the region of inflammation. Original magnification is × 200 and × 40 for A and B, respectively. Taken and modified [16].

MHV MODEL

Mouse Hepatitis Virus or MHV was first isolated in 1949 from a mouse brain showing symptoms of paralysis with dispersed encephalomyelitis and extensive myelin destruction [32]. Later, MHV-induced demyelination was established as a significant experimental animal model to study MS.

MHV is a member of the coronavirus family, having a single-stranded positive-sense RNA genome [33]. MHV strains differ in their cellular tropism and pathogenicity. Based on their organ specificity, MHV strains could be hepatotropic (such as MHV-2), neurotropic (JHM strain), or hepato-neurotropic such as MHV-A59 and MHV-2 [26].

Two commonly used MHV strains for experimental studies are the strongly neuropathogenic JHM strain and the less neurovirulent MHV-A59 strain. Intracranial or intranasal inoculation of mice with MHV results in a biphasic disease condition characterized by acute encephalitis and chronic demyelination [34]. Meningitis (inflammation of meningeal coverings) and encephalomyelitis (accumulation of inflammatory cells in the brain parenchyma) with or without inflammation of the liver (*i.e.,* hepatitis) is observed Fig. (**2**) during the acute infection. The virus reaches its peak titer value within 3-5 days post-infection (p.i). The clearance of infectious viral particles takes place within 10-15 days p.i, and mice start to develop demyelination accompanied by hind limb weakness [26].

The viral entry begins with the interaction between the viral envelope protein or spike protein (S) and its cellular host receptor. The receptor-spike interaction starts the fusion of the viral and plasma membranes with subsequent release of the viral genome in the cytoplasm [33].

The primary cellular receptor for MHV is CEACAM1a. The cellular adhesion molecule, CEACAM1a, is a member of the carcinoembryonic antigen family and belongs to the superfamily of immunoglobulins. CEACAM1a plays a diverse role in normal physiology involving adhesion, suppression of tumors, and immune signaling [34, 35].

There is a high expression of CEACAM1a in epithelial cells, macrophages, T cells, and B cells. However, the expression of CEACAM1a in the brain is restricted only to microglia and the endothelial cells lining the brain microvessels. In contrast, MHV can infect different types of CNS cells, having a poor expression of the CEACAM1a receptor. This observation suggests the existence of other receptors or alternate methods of MHV infection [34].

MHV MEDIATED IMMUNE RESPONSES

Type I interferon (IFN-α/β) mediated host defense mechanism plays a vital role in the early acute stage of infection with MHV strains. Studies have shown that mice deficient in type I interferon receptor (IFNAR−/−) upon intracranial infection with low doses of neurotropic MHV-A59 or JHM.SD strains have higher levels of viral titer in the CNS and also have a higher mortality rate than wild-type mice [36].

The innate immune reaction following MHV infection is mediated by the migration of macrophages, neutrophils, and natural killer (NK) cells from the periphery into the CNS and the secretion of inflammatory chemokines. Depending on the MHV strain, the infiltrating cell types and the nature of secretory chemokines vary [33]. Additionally, recent studies have demonstrated the critical role played by T-cells during acute infection. Adequate clearance of the infectious virus requires CD^{8+} T cells. Viral titer reaches peak levels at 5 days post-infection, and the subsequent decline in viral titer coincides with the build-up of CD^{8+} T cells specific against the virus in the CNS [33]. Over time, the number of CD^{8+} lymphocytes decreases but is not eliminated from the CNS completely. Moreover, viral RNA persists throughout the life-span of mice, even when no viral particles are detectable in the CNS [33].

Furthermore, the function of CD^{8+} cells is dependent on CD^{4+} T cells, as evident from the observation that MHV-specific CD^{8+} cells, when transferred in mice lacking CD^{4+} cells, do not result in significant clearance of viral particles when compared with wild-type mice [37]. Surprisingly, CD^{4+} T cells also have a role in MHV-induced pathogenicity. The epitope M133 from a neurovirulent strain of JHM when mutated resulted in decreased mortality in C57BL/6 mice while it did not affect mortality in BALB/c mice that do not recognize M133. The fact that M133 is presented *via* MHC class-II to CD^{4+} lymphocytes [38] is suggestive of the dual role played by T lymphocytes during infection.

Additionally, the anti-inflammatory role of regulatory CD^{4+} T cells (Treg) has gained importance [39]. Studies involving the transfer of Treg cells in MHV-infected wild-type C57BL/6 mice or RAG1-/- mice (lacks both T and B cells) have shown that there is better survival, reduced accumulation of inflammatory cells, and less demyelination upon transfer of Tregs [40, 41]. On the other hand, prior depletion of Tregs resulted in severe demyelination upon infection with MHV [42]. It has been reported that Tregs remains in the draining cervical lymph nodes, wherein it suppresses self-reactive CD^{4+} T cells [39].

Furthermore, the ability to control pro-inflammatory responses has also been demonstrated in CD^{8+} T cells expressing IL-10 cytokine. Mice deficient in IL-10, when infected with attenuated JHMV strain, suffers from severe disease outcomes with increased infiltration of inflammatory cells and higher mortality [39]. IL-10 expression by CD^{8+} cells is transient and falls as the inflammation diminishes. Studies involving neutralization of IL-10 at different stages post-infection in mice showed that when neutralized at an early stage (days 1-11) compared to later period, it results in increased demyelination, suggesting that IL-10 is required during the early infection stage [43]. Also, IL-10 also helps in tissue repair. Activation of astrocytes and glial scar formation, both of which limits myelin damage, gets compromised in the absence of IL-10 [44].

Thus, Tregs, with its ability to reduce demyelination and suppress the proliferation of auto-reactive T cells, and IL, with its anti-inflammatory property, are effective in treating virus-induced demyelination in the experimental model and may be useful in treating persons infected with viruses causing demyelination [39].

Comparative studies involving two glial-tropic variants of JHM strain that differs in their CNS persistence ability revealed that the continued presence of viral RNA is necessary to maintain the CNS levels of CD4 and CD8 T cells, suggesting that viral persistence provides a signal that maintains lymphocyte populations within the CNS [45].

Finally, the humoral response plays a protective function during MHV infection. Neutralizing antibodies (Abs) appear in the serum once the viral particles are removed and stays elevated, suggesting that virus-specific Abs prevent the re-emergence of infectious virus [33]. Altogether, T cell-mediated cytolysis and antibody production collectively controls MHV persistence in the CNS.

Recombinant Demyelinating (DM) And Non-Demyelinating (NDM) Mhv Strains

CNS infection with mouse hepatitis virus (MHV) is a well-established animal model to study diseases associated with demyelination. Infection with highly neurovirulent strains of MHV, such as JHM, usually results in the death of infected mice at an early stage from severe encephalitis with few or no survivors, thus making it difficult to study the late demyelinating phase. Less neurovirulent MHV-A59 strains are therefore preferred to study demyelination. MHV-A59 infection induces demyelination in the absence of detectable viral particles in the CNS [26], which is in contrast to TMEV-induced demyelination, in which there is the persistence of the infectious virus. Fig. (**3**) shows the CNS pathology of EGFP expressing recombinant MHV strains.

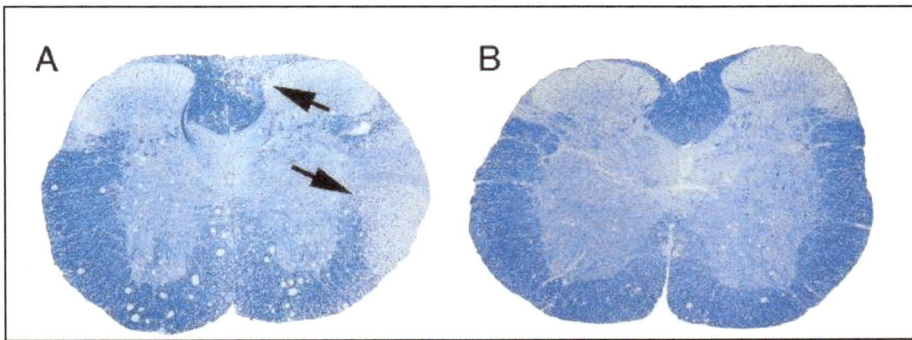

Fig. (3). CNS pathology of EGFP expressing recombinant MHV strains. Paraffin-embedded spinal cord sections taken from mice infected with RSA59 (**A**) or RSMHV2 (**B**) at day 30 post-infection and stained with luxol fast blue and cresyl violet. Regions of demyelination (arrows) evident in the RSA59 infected tissue section. Myelin remains intact upon RSMHV2 infection. Original magnification is ×40. Taken from [50].

Two naturally occurring MHV strains, MHV-A59 [46], and MHV-2 [47] differ in their ability to induce demyelination. MHV-A59 causes demyelination, while MHV-2, closely related to MHV-A59, fails to do so. Genomic sequence comparison between MHV-2 and MHV-A59 revealed a significant difference in the spike gene sequence. There are 84% common amino acid residues with additional 10% similar residues in the spike (S) protein between MHV-A59 and MHV-2. However, 6% of residues in the S protein have no similarity between the two strains [26].

Targeted RNA recombination technology has helped in the construction of new MHV strains. These recombinant strains are labeled RSA59 and RSMHV2. Both are isogenic strains having the same MHV-A59 genomic background. They only differ in their spike gene. RSA59 carries the MHV-A59 spike, and RSMHV2 bears the MHV-2 spike. These recombinant strains also express EGFP (enhanced green fluorescent protein) to aid in the detection and tracking of the virus both *in vivo* and *in vitro*, respectively [48]. Routine histopathological examination demonstrated that the recombinant strains differ in their ability to induce demyelination. While RSA59 (DM strain) infections result in demyelination, RSMHV2 fails to do so (Fig. **4**) . Due to its inability to induce demyelination, RSMHV2 (NDM strain) serves as an appropriate negative control in experimental demyelination studies. Fig. (**3**) shows the demonstration of myelin and axonal preservation upon infection with recombinant MHV strains.

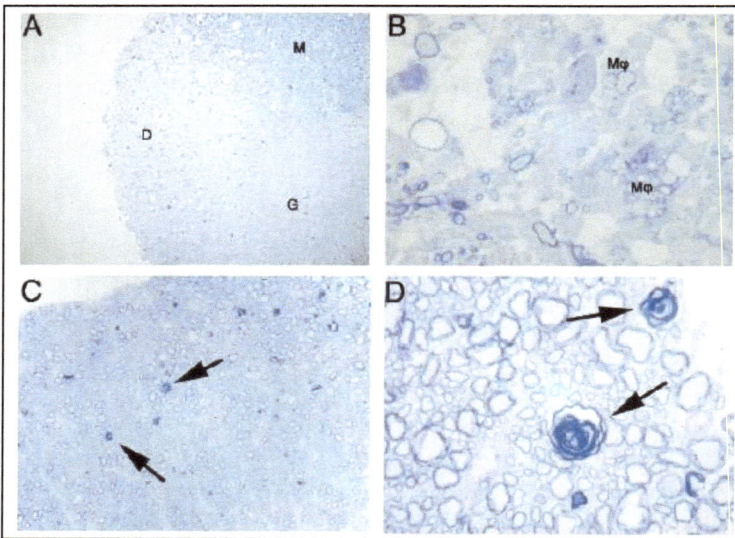

Fig. (**4**). Demonstration of myelin and axonal preservation upon infection with recombinant MHV strains. 1 µm thick spinal cord sections from mice sacrificed at day 30 post-infection with RSA59 (DM) and RSMHV2 (NDM) were stained axonal detection with toluidine blue. Large demyelinating plaques are evident in DM-infected spinal cord sections (**A, B**). In contrast, NDM infection (**C, D**) relatively preserves the myelin covering with rare evidence of early-stage axonal damage (arrows). Marked are myelinated white matter (M), demyelinated white matter region (**D**), grey matter (**G**), and macrophages (Mφ). Original magnificationis 100× (**A and C**) and 1000× (**B and D**). Taken from [50].

Evidence from recent studies suggests that infection of mice with MHV-A59 or RSA59 results in optic neuritis characterized by simultaneous axonal loss and demyelination [49]. Additionally, RSA59 infection results in substantial loss of axonal staining in areas of demyelination in spinal cord sections. In contrast, RSMHV2 infected spinal cords show no loss in axonal staining, suggesting that RSA59 infection induces concurrent axonal loss with demyelination (Fig. **5**) [50]. These findings are in favor of the inside-out model of axonal damage, where the primary targets are axons, followed by the destruction of the myelin sheath.

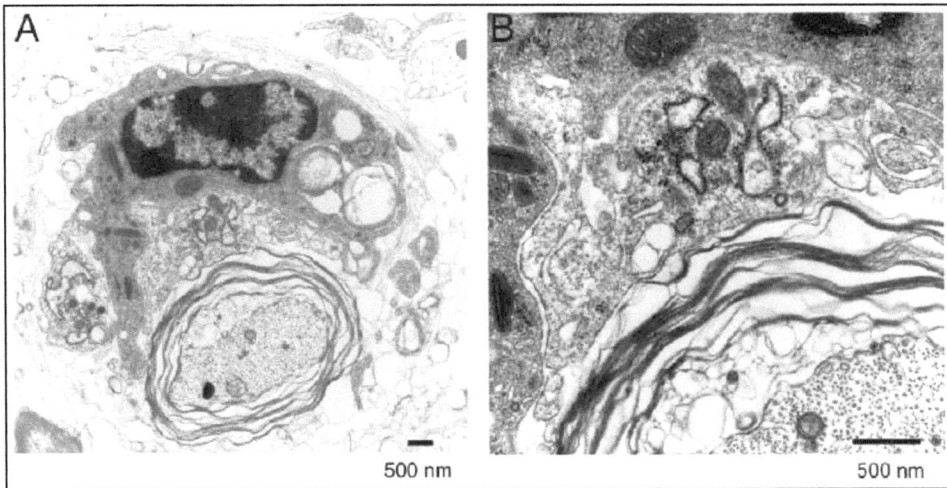

Fig. (5).   Macrophage or microglia-mediated myelin stripping upon RSA59 infection in mice. High-resolution TEM images revealed close localization of a macrophage (top portion of figure A) to a neighboring intact axon (bottom part of figure A) with indications of myelin unraveling. Myelin fragments seen within the cytoplasm of this macrophage are indicative of myelin engulfment. Higher magnification (B) revealed macrophage cell membrane in close contact with uncompacted myelin layers. Multiple vacuoles are present at the inner layers of the myelin covering due to the lifting of myelin from the axon. Original magnification is ×15,000 and ×40,000 for A and B, respectively. Taken from [50].

Recently, it has been demonstrated that the viral spike protein plays a critical role in mediating axonal transportation of the virus. Intracranial infection of mice with the NDM strain fails to extend inflammation into the spinal cord white matter due to a lack of viral antigen spread. In contrast, infection with DM strain results in extensive white matter pathology facilitated by microglia or macrophages [50]. Fig. (**4**) shows macrophage or microglia-mediated myelin stripping upon RSA59 infection in mice.

The persistence of viral RNA during chronic disease appears to be a deciding factor in MHV-induced demyelination. MHV strains, such as MHV-2 that do not persist, fails to produce demyelination. However, the observation that recombinant laboratory strains of MHV such as Penn 98-1 and Penn 98-2, though continuous, are demyelination-negative is suggestive that viral persistence cannot be the only deciding factor to induce demyelination. Instead, the persistence of the virus in specific cell types appears to be critical for demyelination, as evident from the observation that Penn 98-1 and Penn 98-2 may persist in the grey matter while demyelinating strains (MHV-A59 or RSA59) exists in the white matter region [51]. The individual function of different CNS cells in MHV-induced demyelination demands further exploration.

Although the exact mechanism of demyelination induced by MHV is unclear, enough evidence exists to support immune-mediated myelin destruction as one of the primary modes of the demyelination process [50, 52]. A recent study demonstrated the presence of macrophages (derived from monocytes)or CNS resident microglia in areas of demyelination and their adjacent localization to axons upon infection with RSA59 (Fig. **4**). These data suggest macrophage/microglia-mediated myelin stripping during chronic inflammatory demyelination [50]. Further, infection of Rag-1 knockout mice with MHV-JHM resulted in few lesions in mice at 7–15 days post-infection regardless of viral replication, indicating the requirement of lymphocytes during chronic demyelination [52].

CONCLUDING REMARKS

MS is a complex disease with a multitude of factors responsible for the disease onset. Limitations of human studies, as described in the chapter, further makes it difficult to identify triggers of MS. Therefore, appropriate experimental animal models are needed to understand MS pathogenesis and new therapeutic challenges better. There is no one experimental animal model that covers the entire gamut of the disease. The availability of a broad spectrum of different models allows studying individual aspects and mechanisms relevant to disease pathogenesis. The selection of the right model largely depends upon the specific question to be addressed. The EAE model has helped in delineating the role of adaptive immune effectors in demyelination. On the other hand, the TMEV-induced demyelination model has demonstrated molecular mimicry and epitope spreading as the underlying mechanisms of virus-induced autoimmunity in the central nervous system. The diversity of neurotropic strains, in combination with genetic manipulation tools, makes MHV a suitable model to study the host and viral factors determining disease outcome. Our understanding of immune-mediated CNS damage has been further enhanced by studies in MHV-induced neuroinflammatory demyelination, suggesting macrophage-mediated myelin stripping in neurodegeneration. Studies in MHV have also expanded our understanding of axonal damage, demonstrating that axonal injury coincides with demyelination. A significant advantage of viral models is that inflammation is induced by an infectious agent, thus providing a more natural means to understand the etiology of the disease. Studies in animal models are indispensable in the testing of novel therapies and recognizing the physiological side-effects of newly discovered drugs.

Proof in support of viruses playing a role in the pathogenesis of MS is indirect because no specific virus causing MS is detected so far. Further, extensive investigations of these models are needed to understand how the presence of viral particles at low levels can induce chronic neuroinflammation. Additionally, the role of the innate and adaptive immune system during chronic inflammation needs to be explored. Further investigations are needed to address how viruses enter and spread in cells exhibiting low levels of the receptor. Insights into the process by which viruses initiate demyelination in an animal model have crucial applications in the cure of demyelinating disease in humans.

Besides, there is a need to follow strict rules of experimentation to achieve reproducible results. This medium has gained importance because published preliminary reports have often failed to reproduce comparable results when clinically tested in patients. Recently, standard guidelines have been set for researches involving the application of grants and publications in EAE [53]. Replication of a similar approach in other models described in this chapter is appreciable.

CONSENT FOR PUBLICATION

Not applicable.

CONFLICT OF INTEREST

The authors declare no conflict of interest, financial or otherwise.

ACKNOWLEDGEMENTS

The present study has been supported by the Indian Institute of Science Education and Research Kolkata (IISER-K), India. The accomplishment of the current book chapter is derived from the work of researchers whose names are mentioned in the references.

REFERENCES

[1]   Kempuraj D, Thangavel R, Natteru PA, *et al.* Neuroinflammation induces neurodegeneration. J Neurol Neurosurg Spine 2016; 1(1): 1003.
[PMID: 28127589]

[2]   Singh M, Das Sarma J. Demyelinating diseases and neuroinflammation. In: Jana N, Basu A, Tandon

PN, Eds. Inflammation: the Common Link in Brain Pathologies. Singapore: Springer Singapore 2016; pp. 139-70.
[http://dx.doi.org/10.1007/978-981-10-1711-7_5]

[3]   National Multiple Sclerosis Society. What is MS? 2018 [Available from: https://www.nationalmssociety.org/What-is-MS.

[4]   Frischer JM, Bramow S, Dal-Bianco A, *et al.* The relation between inflammation and neurodegeneration in multiple sclerosis brains. Brain 2009; 132(Pt 5): 1175-89.
[http://dx.doi.org/10.1093/brain/awp070] [PMID: 19339255]

[5]   Frohman EM, Racke MK, Raine CS. Multiple sclerosis-the plaque and its pathogenesis. N Engl J Med 2006; 354(9): 942-55.
[http://dx.doi.org/10.1056/NEJMra052130] [PMID: 16510748]

[6]   Scalfari A, Neuhaus A, Daumer M, Ebers GC, Muraro PA. Age and disability accumulation in multiple sclerosis. Neurology 2011; 77(13): 1246-52.
[http://dx.doi.org/10.1212/WNL.0b013e318230a17d] [PMID: 21917763]

[7]   Sospedra M, Martin R. Immunology of multiple sclerosis. Annu Rev Immunol 2005; 23: 683-747.
[http://dx.doi.org/10.1146/annurev.immunol.23.021704.115707] [PMID: 15771584]

[8]   Ascherio A, Munger KL. Environmental risk factors for multiple sclerosis. Part I: the role of infection. Ann Neurol 2007; 61(4): 288-99.
[http://dx.doi.org/10.1002/ana.21117] [PMID: 17444504]

[9]   Ascherio A, Munger KL. Environmental risk factors for multiple sclerosis. Part II: Noninfectious factors. Ann Neurol 2007; 61(6): 504-13.
[http://dx.doi.org/10.1002/ana.21141] [PMID: 17492755]

[10]   Oleszak EL, Chang JR, Friedman H, Katsetos CD, Platsoucas CD. Theiler's virus infection: a model for multiple sclerosis. Clin Microbiol Rev 2004; 17(1): 174-207.
[http://dx.doi.org/10.1128/CMR.17.1.174-207.2004] [PMID: 14726460]

[11]   Kular L, Liu Y, Ruhrmann S, *et al.* DNA methylation as a mediator of HLA-DRB1*15:01 and a protective variant in multiple sclerosis. Nat Commun 2018; 9(1): 2397.
[http://dx.doi.org/10.1038/s41467-018-04732-5] [PMID: 29921915]

[12]   Wandinger K, Jabs W, Siekhaus A, *et al.* Association between clinical disease activity and Epstein-Barr virus reactivation in MS. Neurology 2000; 55(2): 178-84.
[http://dx.doi.org/10.1212/WNL.55.2.178] [PMID: 10908887]

[13]   Moore FG, Wolfson C. Human herpes virus 6 and multiple sclerosis. Acta Neurol Scand 2002; 106(2): 63-83.
[http://dx.doi.org/10.1034/j.1600-0404.2002.01251.x] [PMID: 12100366]

[14]   Gilden DH. Infectious causes of multiple sclerosis. Lancet Neurol 2005; 4(3): 195-202.
[http://dx.doi.org/10.1016/S1474-4422(05)70023-5] [PMID: 15721830]

[15]   Connolly JH, Allen IV, Hurwitz LJ, Millar JH. Measles-virus antibody and antigen in subacute sclerosing panencephalitis. Lancet 1967; 1(7489): 542-4.
[http://dx.doi.org/10.1016/S0140-6736(67)92117-4] [PMID: 4163906]

[16]   Chatterjee D, Biswas K, Nag S, Ramachandra SG, Das Sarma J. Microglia play a major role in direct viral-induced demyelination. Clin Dev Immunol 2013; 2013: 510396.
[http://dx.doi.org/10.1155/2013/510396] [PMID: 23864878]

[17]   Gandhi R, Laroni A, Weiner HL. Role of the innate immune system in the pathogenesis of multiple sclerosis. J Neuroimmunol 2010; 221(1-2): 7-14.
[http://dx.doi.org/10.1016/j.jneuroim.2009.10.015] [PMID: 19931190]

[18]   Costello KHJ, Kalb R, Skutnik L, Rapp R. The use of disease-modifying therpaies in multiple sclerosis: principles and current evidence. A consensus paper by Multiple Sclerosis Coalition 2014.

Available from: http://www.nationalmssociety.org /getmedia/5ca284d3-fc7c-4ba5-b005-ab537d4 95c3c/DMT_Consens us_MS_Coalition_color.

[19]　Rivers TM, Sprunt DH, Berry GP. Observations on attempts to produce acute disseminated encephalomyelitis in monkeys. J Exp Med 1933; 58(1): 39-53.
[http://dx.doi.org/10.1084/jem.58.1.39] [PMID: 19870180]

[20]　Rivers TM, Schwentker FF. Encephalomyelitis accompanied by myelin destruction experimentally produced in monkeys. J Exp Med 1935; 61(5): 689-702.
[http://dx.doi.org/10.1084/jem.61.5.689] [PMID: 19870385]

[21]　Gold R, Linington C, Lassmann H. Understanding pathogenesis and therapy of multiple sclerosis *via* animal models: 70 years of merits and culprits in experimental autoimmune encephalomyelitis research. Brain 2006; 129(Pt 8): 1953-71.
[http://dx.doi.org/10.1093/brain/awl075] [PMID: 16632554]

[22]　Kuerten S, Gruppe TL, Laurentius LM, *et al.* Differential patterns of spinal cord pathology induced by MP4, MOG peptide 35-55, and PLP peptide 178-191 in C57BL/6 mice. APMIS 2011; 119(6): 336-46.
[http://dx.doi.org/10.1111/j.1600-0463.2011.02744.x] [PMID: 21569091]

[23]　Kuerten S, Javeri S, Tary-Lehmann M, Lehmann PV, Angelov DN. Fundamental differences in the dynamics of CNS lesion development and composition in MP4- and MOG peptide 35-55-induced experimental autoimmune encephalomyelitis. Clin Immunol 2008; 129(2): 256-67.
[http://dx.doi.org/10.1016/j.clim.2008.07.016] [PMID: 18722816]

[24]　Glatigny S, Bettelli E. Experimental autoimmune encephalomyelitis (EAE) as animal models of multiple sclerosis (MS). Cold Spring Harb Perspect Med 2018; 8(11): a028977.
[http://dx.doi.org/10.1101/cshperspect.a028977] [PMID: 29311122]

[25]　Jäger A, Dardalhon V, Sobel RA, Bettelli E, Kuchroo VK. Th1, Th17, and Th9 effector cells induce experimental autoimmune encephalomyelitis with different pathological phenotypes. J Immunol 2009; 183(11): 7169-77.
[http://dx.doi.org/10.4049/jimmunol.0901906] [PMID: 19890056]

[26]　Das Sarma J. A mechanism of virus-induced demyelination. Interdiscip Perspect Infect Dis 2010; 2010: 109239.
[http://dx.doi.org/10.1155/2010/109239] [PMID: 20652053]

[27]　Stohlman SA, Hinton DR. Viral induced demyelination. Brain Pathol 2001; 11(1): 92-106.
[http://dx.doi.org/10.1111/j.1750-3639.2001.tb00384.x] [PMID: 11145206]

[28]　Theiler M. Spontaneous encephalomyelitis of mice-a new virus disease. Science 1934; 80(2066): 122.
[http://dx.doi.org/10.1126/science.80.2066.122-a] [PMID: 17750712]

[29]　Theiler M. Spontaneous encephalomyelitis of mice, a new virus disease. J Exp Med 1937; 65(5): 705-19.
[http://dx.doi.org/10.1084/jem.65.5.705] [PMID: 19870629]

[30]　Mecha M, Carrillo-Salinas FJ, Mestre L, Feliú A, Guaza C. Viral models of multiple sclerosis: neurodegeneration and demyelination in mice infected with Theiler's virus. Prog Neurobiol 2013; 101-102: 46-64.
[http://dx.doi.org/10.1016/j.pneurobio.2012.11.003] [PMID: 23201558]

[31]　Njenga MK, Coenen MJ, DeCuir N, Yeh HY, Rodriguez M. Short-term treatment with interferon-alpha/beta promotes remyelination, whereas long-term treatment aggravates demyelination in a murine model of multiple sclerosis. J Neurosci Res 2000; 59(5): 661-70.
[http://dx.doi.org/10.1002/(SICI)1097-4547(20000301)59:5<661::AID-JNR9>3.0.CO;2-E] [PMID: 10686594]

[32]　Lassmann H, Bradl M. Multiple sclerosis: experimental models and reality. Acta Neuropathol 2017; 133(2): 223-44.
[http://dx.doi.org/10.1007/s00401-016-1631-4] [PMID: 27766432]

[33]   Bergmann CC, Lane TE, Stohlman SA. Coronavirus infection of the central nervous system: host-virus stand-off. Nat Rev Microbiol 2006; 4(2): 121-32.
[http://dx.doi.org/10.1038/nrmicro1343] [PMID: 16415928]

[34]   Bender SJ, Weiss SR. Pathogenesis of murine coronavirus in the central nervous system. J Neuroimmune Pharmacol 2010; 5(3): 336-54.
[http://dx.doi.org/10.1007/s11481-010-9202-2] [PMID: 20369302]

[35]   Kuespert K, Pils S, Hauck CR. CEACAMs: their role in physiology and pathophysiology. Curr Opin Cell Biol 2006; 18(5): 565-71.
[http://dx.doi.org/10.1016/j.ceb.2006.08.008] [PMID: 16919437]

[36]   Roth-Cross JK, Bender SJ, Weiss SR. Murine coronavirus mouse hepatitis virus is recognized by MDA5 and induces type I interferon in brain macrophages/microglia. J Virol 2008; 82(20): 9829-38.
[http://dx.doi.org/10.1128/JVI.01199-08] [PMID: 18667505]

[37]   Stohlman SA, Bergmann CC, Lin MT, Cua DJ, Hinton DR. CTL effector function within the central nervous system requires CD4^+ T cells. J Immunol 1998; 160(6): 2896-904.
[PMID: 9510193]

[38]   Anghelina D, Pewe L, Perlman S. Pathogenic role for virus-specific CD4 T cells in mice with coronavirus-induced acute encephalitis. Am J Pathol 2006; 169(1): 209-22.
[http://dx.doi.org/10.2353/ajpath.2006.051308] [PMID: 16816374]

[39]   Perlman S, Zhao J. Roles of regulatory T cells and IL-10 in virus-induced demyelination. J Neuroimmunol 2017; 308: 6-11.
[http://dx.doi.org/10.1016/j.jneuroim.2017.01.001] [PMID: 28065579]

[40]   Anghelina D, Zhao J, Trandem K, Perlman S. Role of regulatory T cells in coronavirus-induced acute encephalitis. Virology 2009; 385(2): 358-67.
[http://dx.doi.org/10.1016/j.virol.2008.12.014] [PMID: 19141357]

[41]   Trandem K, Anghelina D, Zhao J, Perlman S. Regulatory T cells inhibit T cell proliferation and decrease demyelination in mice chronically infected with a coronavirus. J Immunol 2010; 184(8): 4391-400.
[http://dx.doi.org/10.4049/jimmunol.0903918] [PMID: 20208000]

[42]   de Aquino MT, Puntambekar SS, Savarin C, *et al.* Role of CD25(+) CD4(+) T cells in acute and persistent coronavirus infection of the central nervous system. Virology 2013; 447(1-2): 112-20.
[http://dx.doi.org/10.1016/j.virol.2013.08.030] [PMID: 24210105]

[43]   Trandem K, Jin Q, Weiss KA, James BR, Zhao J, Perlman S. Virally expressed interleukin-10 ameliorates acute encephalomyelitis and chronic demyelination in coronavirus-infected mice. J Virol 2011; 85(14): 6822-31.
[http://dx.doi.org/10.1128/JVI.00510-11] [PMID: 21593179]

[44]   Puntambekar SS, Hinton DR, Yin X, *et al.* Interleukin-10 is a critical regulator of white matter lesion containment following viral induced demyelination. Glia 2015; 63(11): 2106-20.
[http://dx.doi.org/10.1002/glia.22880] [PMID: 26132901]

[45]   Marten NW, Stohlman SA, Bergmann CC. Role of viral persistence in retaining CD8(+) T cells within the central nervous system. J Virol 2000; 74(17): 7903-10.
[http://dx.doi.org/10.1128/JVI.74.17.7903-7910.2000] [PMID: 10933698]

[46]   Lavi E, Gilden DH, Wroblewska Z, Rorke LB, Weiss SR. Experimental demyelination produced by the A59 strain of mouse hepatitis virus. Neurology 1984; 34(5): 597-603.
[http://dx.doi.org/10.1212/WNL.34.5.597] [PMID: 6324031]

[47]   Keck JG, Soe LH, Makino S, Stohlman SA, Lai MM. RNA recombination of murine coronaviruses: recombination between fusion-positive mouse hepatitis virus A59 and fusion-negative mouse hepatitis virus 2. J Virol 1988; 62(6): 1989-98.
[http://dx.doi.org/10.1128/JVI.62.6.1989-1998.1988] [PMID: 2835504]

[48]   Das Sarma J, Scheen E, Seo SH, Koval M, Weiss SR. Enhanced green fluorescent protein expression may be used to monitor murine coronavirus spread *in vitro* and in the mouse central nervous system. J Neurovirol 2002; 8(5): 381-91.
[http://dx.doi.org/10.1080/13550280260422686] [PMID: 12402164]

[49]   Shindler KS, Kenyon LC, Dutt M, Hingley ST, Das Sarma J. Experimental optic neuritis induced by a demyelinating strain of mouse hepatitis virus. J Virol 2008; 82(17): 8882-6.
[http://dx.doi.org/10.1128/JVI.00920-08] [PMID: 18579591]

[50]   Das Sarma J, Kenyon LC, Hingley ST, Shindler KS. Mechanisms of primary axonal damage in a viral model of multiple sclerosis. J Neurosci 2009; 29(33): 10272-80.
[http://dx.doi.org/10.1523/JNEUROSCI.1975-09.2009] [PMID: 19692601]

[51]   Das Sarma J, Fu L, Tsai JC, Weiss SR, Lavi E. Demyelination determinants map to the spike glycoprotein gene of coronavirus mouse hepatitis virus. J Virol 2000; 74(19): 9206-13.
[http://dx.doi.org/10.1128/JVI.74.19.9206-9213.2000] [PMID: 10982367]

[52]   Wu GF, Dandekar AA, Pewe L, Perlman S. CD4 and CD8 T cells have redundant but not identical roles in virus-induced demyelination. J Immunol 2000; 165(4): 2278-86.
[http://dx.doi.org/10.4049/jimmunol.165.4.2278] [PMID: 10925317]

[53]   Baker D, Amor S. Publication guidelines for refereeing and reporting on animal use in experimental autoimmune encephalomyelitis. J Neuroimmunol 2012; 242(1-2): 78-83.
[http://dx.doi.org/10.1016/j.jneuroim.2011.11.003] [PMID: 22119102]

CHAPTER 8

Novel Therapeutic Targets in Amyotrophic Lateral Sclerosis

Priyanka Gautam, Mukesh Kumar Jogi and **Abhishek Pathak**[*]

Department of Neurology, Institute of Medical Sciences, Banaras Hindu University, Varanasi, Uttar Pradesh, India

Abstract: Amyotrophic Lateral Sclerosis is an adult-onset, irremediable, and fatal neurodegenerative disease marked by the advancement in the loss of motor neurons in the spinal cord, brain stem, and motor cortex. Etiology is blurred, but it is thought to be multifactorial, which contributes to the heterogeneity and complexity of the disease. Core knowledge of primary etiology and pathological mechanisms can pave the way towards treatment. This chapter examines mechanisms that may contribute to motor neuron degeneration, among which oxidative stress, mitochondrial dysfunction, protein aggregation, axonal transport are potential novel therapeutic targets for ALS treatment.

Keywords: Amyotrophic lateral sclerosis, Motor neuron, Neurodegenerative disease.

INTRODUCTION

Amyotrophic lateral sclerosis (ALS), sometimes referred to as "Lou Gehrig's disease," is characterized by progressive deterioration of the upper motor (UMN) and lower motor neurons (LMN), which regulate muscle weakness and eventually leading to paralysis [1]. The initial presentation of ALS may vary among patients; some are present with spinal-onset disease, but others can present with bulbar-onset disease, which are characterized by dysarthria and dysphagia. In most patients, the root of ALS is mysterious; however, some individuals have familial disease, which is associated with changes in genes that have a variety of functions, including roles in non-motor cells. In familial ALS, some of the implicated genes are not fully penetrant, and with rare exceptions, the genotype does not necessarily predict phenotype [2].

Respiratory failure occurs in most patients within 2-5 years after diagnosis due to respiratory muscle involvement. Some patients may have a cognitive impairment,

[*] **Corresponding author Abhishek Pathak:** Department of Neurology, Institute of Medical Sciences, Banaras Hindu University, Varanasi, Uttar Pradesh, India; Tel: +91 8840139503; E-mail: abhishekpathakaiims@gmail.com

Sachchida Nand Rai (Ed.)

further adding to the damage. Only10% of the patients survive with 10 years of lifespan [3]. There has been a gradual increase over the past few years in ALS, although no unified hypothesis has been revealed for its pathogenesis. Most cases of ALS are sporadic (SALS), although some are classified as 'familial ALS' (FALS) because few causal genes have been identified in the last three decades. The study of these gene variants has led to many new pathophysiological concepts and new therapeutic approaches [4].

HISTORY

American baseball player Lou Gehrig was diagnosed with ALS, after which ALS is also known as Lou Gehrig's disease. His disease was under investigation for his immediate loss of performance in baseball, and as a result, there was an early withdrawal, and two years later, he died at the age of 37. Another renowned personality who was a British physicist was Stephen Hawking, diagnosed with ALS at the age of 21, and he lived over 70 years under palliative care [5].

EPIDEMIOLOGY

ALS incidence is 4.1-8.4 per 100,000 people in recent population-based studies [6]. The universal incidence of ALS has increased over the years, especially in Western societies. ALS is an uncommon disease, and its incidence is estimated to be 2-3 per 100,000 in Europe and 0.7–0.8 per 100,000 individuals in Asia [1]. The prevalence of ALS is higher in people aged over 50 years and is 6 per 100,000 of the total population. Only 10% of cases are familial (inherited from parents), and the remaining 90% of cases are sporadic. According to the Foundation for Research on Rare Diseases and Disorders [fRRDD], the frequency of ALS cases in India is 5 per 100,000, and the male to female ratio is 2:1 [5].

SIGN AND SYMPTOMS

Muscle weakness, torsion, and numbness are common symptoms found in both types of ALS that can lead to muscle detriments [7, 8]. ALS patients develop symptoms of dyspnoea and dysphagia at their most advanced stage [9]. The preparative characteristics of ALS are often elusive despite the fatal nature of the disease [10]. Unidentified symptoms delay the diagnosis of ALS, sometimes lead to false diagnosis as well. Retrospective reviews have demonstrated a delay in symptom diagnosis, which has not changed for more than a decade and ranges from 8.0 to 15.6 months [11, 12]. Fig. (1) shows the ALS patient's UL atrophy.

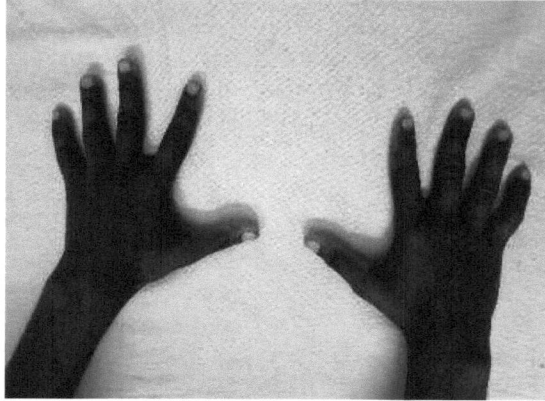

Fig. (1). ALS patients UL atrophy.

CLINICAL CHARACTERISTICS

The clinical diagnosis of classic ALS was found on the detection of progressive dysfunction of cortical UMNs and spinal LMNs in many body regions (mainly, limbs and bulb areas). Much of this presentation is redirected by the El Escorial criteria [13, 14]. In contrast, patients may present with less motor neuron symptoms, including fasciculation, numbness, and muscle wasting. Almost one-third of patients with ALS have bulbar-onset disease, which is characterized by progressive dysarthria following by dysphagia and is sometimes accompanied by emotional dysfunction. Limb-onset disease occurs in about 60% of cases, is usually asymmetrical in appearance, and first develops in the upper or lower limbs. About 5% of patients have respiratory problems, and these patients are often seen in cardiology and pulmonology clinics before being sent to neurology clinics [15]. Fasciculation is a symptom of ALS but are difficult to detect in some patients. Fasciculation arises in the axon of diseased motor neurons in ALS but not specific to the disorder; these include other chronic neurogenic disorders, endocrine/metabolic conditions in some normal people [16 - 18]. The emergence of multidisciplinary, exclusive ALS clinics has enhanced the quality of life and survival of ALS patients, with early non-invasive ventilation as one of the most vital determinants of survival; however, there is still no proper curable treatment for ALS patients [19]. Riluzole is the first FDA approved drug which is available for ALS patients and is believed to act by a variety of mechanisms by blocking the presynaptic release of glutamate, neutralizing voltage-dependent Na^+ channels, reducing hyperactivity, and inactivating K^+ channels, inhibiting kinase C and interfering with exciting transmitter-induced intracellular events; thus exhibiting a modest ability to slow down disease progression, especially when given early in the course of the disease [20].

GENETICS

Evidence from clinical and preliminary research suggests several causes of ALS with important but distinct genetic components. Up to 10% of ALS-affected individuals are members of at least one affected family and are defined as familial ALS (fALS); All these cases are inherited in an autosomal dominant manner [21]. Although more than 50 inducible or disease-modifying genes have been identified, pathogenic variants in SOD1, C9ORF72, FUS, and TARDBP are most common with this disease, while other variants of genes are not common [22]. Studies of the genes, ancestry, and genetic diversity of affected individuals revealed that DNA variation in the C9ORF72, TARDBP, FUS, and SOD1 genes represent 70% of all familial ALS cases [23]. Most of the genetic risk for ALS remains unclear; this means that most research is focused on understanding how alterations and differences in the expression of genes associated with ALS lead to disease. SOD1, TARDBP, FUS, and C9ORF72 exhibited the largest levels [24].

MUTATION

Early-onset SALS patients (onset age <35 years) should be screened for mutations in the FUS gene (especially in patients with bulbar symptoms and rapid progression of the disease), but mutations in the C9ORF72 and SOD1 genes are most likely to be responsible for familial and late-onset sporadic ALS [25]. The discovery of the implication of SOD1 in ALS has greatly accelerated the understanding of the disease, particularly through the development of SOD1 transgenic animal models of ALS (mice and rats). Mutated SOD1 induces neurotoxicity, but the detailed molecular mechanisms of this toxicity are still not fully known. Oxidative stress and iron metabolism disturbances are found in ALS patients carrying SOD1 mutations and SOD1 animal models [26].

In these cases of mutation, SOD1 contains an exposed N-terminal short region, the Darlin-1-binding region (DBR), which induces endoplasmic reticulum stress; It has been suggested that the pathogenic SOD1 mutation induces a normal morphological change in SOD1 that can lead to motor neuron toxicity [27]. The gain of function in the C9ORF72 gene can lead to repeats of hexanucleotides in ALS patients and affects various downstream pathways, such as DNA damage, dysfunctional nucleolus, nucleo-cytoplasmic transport deficit, endoplasmic reticulum stress, autophagy dysfunction, inhibition of translation & proteasomes, and alteration in dynamics of stress granules. There may be goals in different pathways [28].

ALS PATHOGENIC PATHWAYS AS THERAPEUTIC TARGETS

In this chapter, we will discuss the current therapeutic strategies and various novel targets that can be useful in the treatment of ALS.Molecular mechanisms behind ALS are as follows Oxidative damage, Glutamate Excitotoxicity, Mitochondrial dysfunction, Impaired Axonal Transport, Apoptotic cell death, Glial cell pathology, and Abnormal RNA Metabolism. These molecular mechanisms play a pivotal role in neuronal death in ALS, and therefore, these can be promising potential therapeutic targets for ALS treatment. The following Fig. (**2**) represents different pathways and targets involved in ALS pathology.

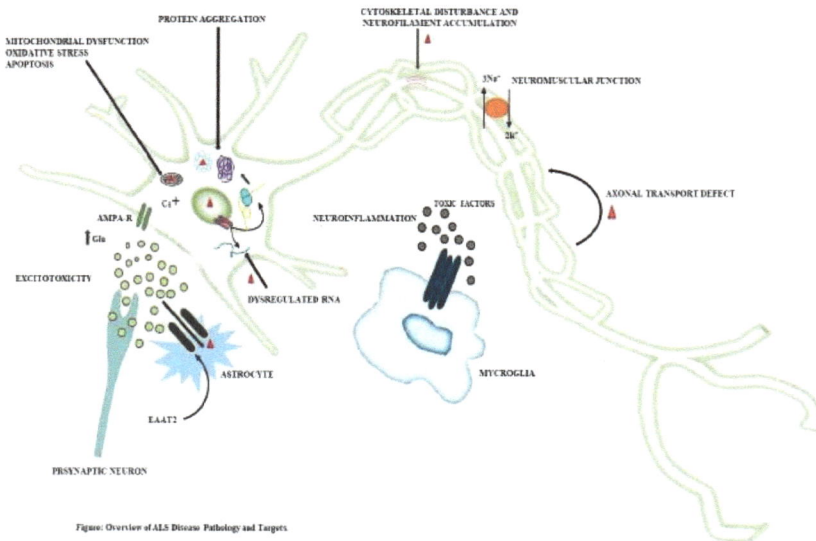

Figure: Overview of ALS Disease Pathology and Targets

Fig. (2). Different pathways and targets involved in ALS pathology.

TARGETING OXIDATIVE STRESS

Oxidative stress is a disparity between the making of free radicals and the ability of the body to purify their damaging effects through neutralization by antioxidants.The cellular antioxidant is encoded by the SOD1 gene, and the mutant form of it leads to the hypothesis of oxidative stress [29]. SOD1-mutations result in oxidative stress in SALS and FALS cases. In the absence of any efficient clinical trials and laboratory studies, oxidative stress is said to play a tributary role in motor neuron degeneration, probably by inducing other pathogenic mechanisms such as mitochondrial dysfunction, glutamate excitotoxicity, and RNA metabolism [30]. No symptoms of ALS or MND are noticed in *SOD1*-knockout mice, which indicates the role of SOD1 in the generation of ALS disease [31].

Table 1. Molecular mechanisms and defectsin amyotrophic lateral sclerosis.

Molecular Mechanism	Associated Alterations
Glutamate Excitotoxicity	Overstimulated EAAT2 Receptor
Protein Aggregation	Impaired Proteostasis
Mitochondrial Dysfunction	Generation of ROS
Apoptosis	Increased level of apoptotic molecules
Neuroinflammation	Production of Toxic Factors
Axonal Transport	Leaching of Cytoskeletal proteins
Altered RNA process	Mislocalization of RNA-binding proteins
Neuromuscular Junction	Pump Dysfunction

Generation of ROS like O2-, H2O2, OH- is because of oxidative phosphorylation inside the cell. Peroxynitrite is the product of $O_2$with NO. These ROS are capable of damaging the cells. Activation of apoptosis occur inside such cells and avoid largescale damage. Mitochondria carry out its function of regulating apoptosis. On binding of mutant mitochondrial SOD1 product with anti-apoptotic protein Bcl-2, functional discrepancy is noticed in mitochondria [32]. The disease advancement in ALS patients can be slowed down by targeting mitochondrial dysfunction.

TARGETING GLUTAMATE TRANSPORT

Production of glutamate in excess causes excitotoxicity, which is to be blamed for minor and major neurodegenerative diseases, including ALS. Astrocytes have a close connection with motor neurons, and they protect the neurons from excitotoxicity. Inflammatory cytokines are released by astrocytes lead to glutamate excitotoxicity, eventually leading to death.

Excitotoxicity has been proved to be an open reason for ALS according to various research outcomes. The excitotoxic potential of excitatory amino acid transporters(EAAT) concluded designing of different subtypes with specific antibodies, expression systems, and probs to delete/knock-down expression, contributing to normal physiology. Regulation of EAA subtypes provided a direction to several studies [33].

Glutamate mediated reciprocity being the reason behind ALS where motor neuron eventually ends up in death. Loss in the activity of glial glutamate transporter EAAT2 results in the accumulation of an acidic amino acid, Glutamate, in the synaptic cleft, which overstimulates the glutamate receptor. Reuptake of glutamate is also forbidden [34]. This makes way for an excessive influx of calcium ions, which causes the death of neuronal cells as a result of disruption of mitochondria and lipid peroxidation. It had been found that excess glutamate in the cerebrospinal fluid of the brain was the cause of excitotoxicity in the patients having ALS [35].

There are several ways to control glutamate excitotoxicity. The major solution is by controlling or inhibiting the release of glutamate, increasing the glutamate transporter function, or blocking the channels for calcium. But, unfortunately, the exact treatment for ALS is still not clear.

There are different drugs targeting excitotoxicity under trial. Ceftriaxone is one among many undergoing clinical trials targeting excitotoxicity. It stimulates the activity of EAAT2 and prevents neuronal death through nuclear factor kappa B(NF-ƙB). It also enhances the uptake of excess glutamate released [36].

Riluzole is the only European Medical Agency (EMA) and the Food and Drug Administration (FDA) approved drugs used in the treatment of ALS as it contains antiglutamate agents. It proved to delay the progression, and survival was improved [37].

TARGETING PROTEIN AGGREGATION

ALS-linked ubiquitous aggregation of proteins like FUS, TDP-43, OPTN, UBQLN2, ATXN2, and proteins from C9ORF72 have been of keen interest and are identified in motor neurons. Different molecular mechanisms lead to protein aggregation, such as low complexity (LC) domains in ALS proteins, stress granule formation, protein sequestering, and non-functionality of protein degradation pathways. Alteration in the process of autophagy and proteasomal degradation form aggregate. Also, definite genetic changes are considered to be responsible for molecular characteristics and distinct protein aggregates. The exact impact of protein aggregation in ALS cannot be explained, but it affects some functionalmechanisms [38]. Proteins that contain low complexity part thatis flexible in shape and different from other proteins are mutated. These parts of proteins form aggregates and affect the other healthy, normally functioning proteins as well. Some molecular components or proteins, which are a type of RNA binding protein (RBP) affect the protein structure. The research for understanding the reason and behavior behind this has been conducted with the help of microscopy, NMR (Nuclear magnetic resonance), spectroscopy, and other computer simulations. There have been comparisons of physical and chemical interactions in healthy and disease-affected proteins. It was found that small changes in the atomic level lead to the formation of protein aggregates. These aggregates can eventually lead to neuronal cell death, ultimately causing ALS [39].

Arimoclomol, a derivative of hydroxylamine, has shown a positive effect in the symptomatic treatment and muscle function improvement of SOD^{G93A}mice but could not increase the lifespan significantly. It increases the expression of Hsp70, which reduces ubiquitin-positive aggregates in treated SOD^{G93A}. This implies that targeting protein aggregation or factors or molecular mechanisms linked to protein aggregation can be explored for the treatment of ALS by reducing protein accumulation [40].

However, an exact phenomenon and procedure to control the situation from arising and understand the relationship between protein aggregation and motor neurons are still under progress.

TARGETING MITOCHONDRIAL DYSFUNCTION

Mitochondrial dysfunction is an initial effect of ALS. It is considered the main disease component for ALS. Accumulation of mutant proteins in mitochondria results in mitochondrial damage and is responsible for ALS [41].

Various important functions of the body like ATP generation, metabolism, lipid biosynthesis, and apoptosis are performed by the mitochondria. It is extremely important for the survival of the neurons and their function. Mitochondrial disturbances make neurons more prone to oxidative stress. The dysfunction leads to motor neuron diseases like ALS. Patients who have ALS have been found with abnormal mitochondria in their motor neurons.

ALS further disrupts mitochondrial functions like ATP production, calcium homeostasis, mitochondrial respiration, and more. Studies have shown damaged DNA in mitochondria of patients with ALS. The dysfunction of the Mitochondrial eventually leads to the death of the neuronal cells. The deficiency of cytochrome c oxidase enzyme has also been observed in samples of patients with ALS [42].

Many bioenergetic agents aim at fighting /reducing mitochondrial dysfunction. Oral administration of creatine in adult patients with abnormal mitochondria behavior was tried to treat the condition. An initial dose of 5 grams, twice a day for two weeks and followed by a dose of a gram, twice a day for 1 week resulted in better anaerobic and aerobic power in the patients. The results with creatine were observed to be better than those with riluzole. But, unfortunately, the two RCTs of creatine did not prove to be successful to show the desired effects in the patient [43, 44]. Dextromethorphan, along with quinidine has shown efficacy in the treatment of ALS. This combination has proven to stabilize the damaged cells and reduced neuronal cell deaths [45].

TARGETING APOPTOSIS

Sometimes cells in the absence or low concentration of accurate signals die by a process known as apoptosis. This is a phenomenon important in the nervous system and brain development. Apoptosis is a sequential process of death of non-functional cells.

The research found that apoptosis could play a role in the process of diseases like ALS. It is suspected that in the initial stage of ALS, apoptosis is triggered. To find the cause and stop the process of motor neuron death due to apoptosis is of concern, as it could lead to therapy and provide treatment of ALS. Apoptotic neuronal death in Sporadic ALS is not evident [46].

There are enzymes present in mitochondria that apoptosis cells, and they can also stop the process. Death receptors, genetic regulation, an inhibitors of apoptosis proteins (IAPs) are said to be responsible for controlling apoptosis. Molecules related to apoptosis have been found in patients with ALS, along with an increased level of BCI-2 (family of oncoprotein) and caspases 1 and 3, which support the involvement of apoptosis in ALS. But, there is no full-proof evidence to state the exact relationship and cause for apoptosis leading to ALS [47].

In *in-vitro* studies, TCH346 was found to be anti-apoptotic, which binds to Glyceraldehyde 3-phosphate Dehydrogenase (GAPDH) and inhibits the apoptotic pathway in which GAPDH is tangled, but the results of the tests were not successful, and it could not show any efficacy on ALS patient [48]. Studies shows, Sodium phenylbutyrate (NaPB), a histone deacetylase inhibitor, improves transcription and post-transcriptional pathways, promoting cell survival in a mouse model of motor neuron disease. NaPBupsurges histone acetylation in ALS Tests with sodium phenylbutyrate was safe and tolerable [49].

Phase 2 trials with lithium carbonate are in progress [50].

TARGETING NEUROINFLAMMATION

Neuroinflammation is a cascade of events mediated by the cytokine production in the central nervous system and is mainly characterized by infiltrating immune cells, astrocytes activation and microglia activation, excess production of cytokine, and T lymphocyte infiltration and macrophage infiltration. Neuroinflammation causes neuronal loss in animals as well as humans.

Studies prove that neuroinflammation leads to the development of ALS. Recent advances focusing on stem cell therapy aimed at reducing its effect in ALS.

Several studies have been conducted on rodents for finding therapies to heal neuroinflammation in ALS. But the preclinical results could not be translated into positive clinical attempts. These tests were done on animals at the initial stage or slowly developing stage of the disease. These tests were successful on animals but failed while trying on patients because the disease in humans was diagnosed at a much later stage [51].

Many ongoing trials, along with the understanding of neuroinflammation, help to find solutions and developments in therapeutics in ALS. Findings proved that treatment, drugs vary at the different stages as the disease progresses, and it was important to promote neuroprotective and anti-inflammatory characteristics of immune cells.

As a result of clinical trials and various studies, replacement or modifications in astrocytes, T- lymphocytes, and microglia are found to be promising and successful treatments for ALS.

Several anti-inflammatory drugs have also been tested in the study of neuroinflammation leading to ALS. These drugs have been tried on rodents, which exhibit some increase in the survival of mice. These anti-inflammatory agents were cyclooxygenase-2 inhibitor, celecoxib, and the hematopoietic cytokine and erythropoietin, but when tried on ALS patients, the result of celecoxib was found to be ineffective [52].

TARGETING GLIAL CELLS (MICROGLIA AND ASTROCYTES)

A specialized group of macrophages are known as microglial cells. Microglia resides inside the immune cells of the brain and in the spinal cord. The main function of microglia is to generate immune responses. They are found to exist in two states - active and resting. Based on the studies and tests conducted on ALS mice, it was found that microglia tend to multiply in number with the passing stage of the disease. Different observations have been made based on the morphology in the early stage and end-stage of the disease.

Microglia are tangled with synaptic organization, trophic neuronal support during development, phagocytosis of apoptotic cells in the developing brain, myelin turnover, control of neuronal excitability, phagocytic debris removal as well as brain protection and repair. There are Glial cells that acquire neurotoxic potential and have been found contributing to neuron cell deaths [53].

Motor neuron injury as an effect of microgliosis is a neuropathological hallmark of ALS. There is no direct death of motor neurons due to exogenous $mSOD1G^{93A}$. On the contrary, enhanced release of pro-inflammatory cytokines and free radicals was noticed due to morphological and functional activation of microglia by $mSOD1\ G^{93A}$ or $mSOD1\ G^{85R}$. Activation of microglia using CD14 and TLR pathways for toxicity in motor neurons is needed directly rather than extracellular $mSOD1G^{93A,}$ which is not directly toxic to motoneurons [54]. Glial reactions are complex, having beneficial as well as a negative impact on motor neurons [55].

This link between mSOD1 and innate immunity possibly can open the way to novel therapeutic targets in ALS [54]. Findings have validated that the replacement of disease-associated microglia with normal, healthy microglia also proves to be a promising approach in the treatment procedure of ALS [56].

Genes that are associated with ALS are also found in astrocytes. Astrocytes that show MSOD1 are toxic to motor neurons. Selective blockage of the gene MSOD1 found in astrocytes has been shown to delay the spread of disease in the tests conducted on the mice. Anti-inflammatory therapeutics so far have not shown success on patients. An improved understanding and the latest tools and technology can provide better solutions and options in this aspect [57]. Acting upon macrophage activation may stabilize the neuromuscular junctions and, therefore, can be considered an option for the treatment of ALS [58].

Histamine is said to modulate microglia associated neuroinflammation. Microglial motility is in the presence of $\alpha5\beta1$ integrin signaling, expressing histamine H_4R is activated by binding of histamine, which curbs LPS induced microglial motility. It also controls the proinflammatory cytokine release [59].

TARGETING DYSREGULATED RNA PROCESS

It has been found that dysregulated RNA has also become a major cause for ALS, where altered RNA metabolism and RNA processing act as triggering factors leading to ALS. Reduction in RBP expression (SMN1, FMRP, RBFOX1, Hu senataxin, angiogenin, NOVA), RBP susceptible aggregation (TDP43, FET proteins, hnRNPs, Tia1, MATR3, Ataxin 2), and Expanded RNA and proteins repossess RBP (DM,ALS/FTD, FXTAS, polyQ) are the mechanisms which tend to modify RNA metabolism. These mechanisms involved in the disruption of RBPs result in neurodegenerativediseases [60].

These complex Aggregates start to modify the RNA characteristics/ functionalities like RNA splicing, capping, transport of RNA, which disturbs the pathway of cell functions, and its morphology [61].

Patients with ALS have also been seen with an excess amount of TDP43 (DNA/RNA binding protein), which is used for stabilizing the RNA in the nerve cells. Excess of this protein -TDP43 is found to accumulate in the cells, which causes extreme inflammation and makes the nerve cells vulnerable to death.

RNAi decreases the expression of specific target genes without affecting other genes or processes. Using longer RNA induces improved immune response leading to repression in neuronal cell deaths [62].

Some miRNAs (micro RNAs) found in ALS have been shown to control apoptosis, autophagy of motor neurons. Based on studies, miRNAs are effective in battling ALS.

TARGETING NEUROMUSCULAR JUNCTION

One of the main reasons cited for leading to ALS is a disturbance in Neuromuscular Junctions(NMs). Disruption in muscle activation due to the changed concentration of Na^+ and K^+ in Na^+K^+ channels in the muscle cause damage in the mitochondria, which causes altered calcium homeostasis leading to apoptosis/necroptosis, eventually affecting the motor neurons [63]. Hence, NMJ plays a critical role in understanding the factors and target treatment in ALS.

The most common and studied experiment is carried out on the transgenic mouse in the ALS model. This research has led to various theories and explanations of the loss of Neuromuscular junction affecting ALS.

Riluzole acts as a stabilizer in sodium channels. It also inhibits the production of glutamate and helps in protecting the neurons and their functions.

Edaravone has the same functions as that vitamin C, and vitamin E. Edaravone helps in reducing oxidative stress in brain cells (microglia, glia, neurons) also minimizes the inflammation responses. Edaravone does not provide a clear mechanism in treating ALS but claim to act as a neuroprotector [64].

TARGETING GENE

Gene therapy is considered to be a possible solution to treat ALS if it manages to provide a beneficial protein to save the dying nerve cells [65]. Researchers can redesign the virus for this purpose, and the virus after redesigning is a vector. These viruses carry therapeutic genes.

Since the cause of this defect is the SOD1 protein, coding in the gene for the SOD1 defect may not work for others. Hence, all patients may not receive the benefit from this idea in ALS. But, from the research and studies, the evidence is clear that it might work with the help of gene therapy in treating ALS. Gene replacement/genre editing /gene correction is a promising option to find success. AAV – Adeno Associated Virus, currently is the most commonly practiced and followed procedure gene therapy-related clinical trial for ALS [66].

In non-vector type gene therapy, RNA is targeted for a therapeutic solution to tackle RNA/Protein accumulation.

After studying and conducting tests on several ALS animal models, SYT13 proved to save the motor nerve cells from damage, and SYT13 was suggested as a promising option for gene therapy [67].

CONCLUDING REMARKS

In 2020, we have a gradient of budding therapeutics, which claim to be effective in ALS treatment but fail to prove so in human clinical trials. This may be due to different reasons. There is a possibility that either we are not testing effective drugs or there is a problem with the design of trials. In the absence of any known accurate cause, it is impossible to have a direction towards treatment. Therefore, we primarily need to know the causes of the disease to have an effective therapeutic target.

CONSENT FOR PUBLICATION

Not applicable.

CONFLICT OF INTEREST

The authors declare no conflict of interest, financial or otherwise.

ACKNOWLEDGEMENTS

The authors are thankful to the Department of Neurology, IMS, BHU for providing the departmental facility.

REFERENCES

[1]   Batra G, Jain M, Singh RS, *et al.* Novel therapeutic targets for amyotrophic lateral sclerosis. Indian J Pharmacol 2019; 51(6): 418-25.
[http://dx.doi.org/10.4103/ijp.IJP_823_19] [PMID: 32029967]

[2]   Al-Chalabi A, van den Berg LH, Veldink J. Gene discovery in amyotrophic lateral sclerosis: implications for clinical management. Nat Rev Neurol 2017; 13(2): 96-104.
[http://dx.doi.org/10.1038/nrneurol.2016.182] [PMID: 27982040]

[3]   Dash RP, Babu RJ, Srinivas NR. Two Decades-Long Journey from Riluzole to Edaravone: Revisiting the Clinical Pharmacokinetics of the Only Two Amyotrophic Lateral Sclerosis Therapeutics. Clin Pharmacokinet 2018; 57(11): 1385-98.
[http://dx.doi.org/10.1007/s40262-018-0655-4] [PMID: 29682695]

[4]   Mathis S, Couratier P, Julian A, Vallat JM, Corcia P, Le Masson G. Management and therapeutic perspectives in amyotrophic lateral sclerosis. Expert Rev Neurother 2017; 17(3): 263-76.
[http://dx.doi.org/10.1080/14737175.2016.1227705] [PMID: 27644548]

[5]   Hancock SM, Iftekhar NT, Jampana SC, *et al. Rare diseases and disorders: Research, Resource and Repository for South Asia.* [Last accessed 15 Oct 2019, 6:30 pm]. Available from: http://www.rarediseasesindia.org/als.

[6]   Longinetti E, Fang F. Epidemiology of amyotrophic lateral sclerosis: an update of recent literature. Curr Opin Neurol 2019; 32(5): 771-6.
[http://dx.doi.org/10.1097/WCO.0000000000000730] [PMID: 31361627]

[7]   Goetz CG. Amyotrophic lateral sclerosis: early contributions of Jean-Martin Charcot. Muscle Nerve 2000; 23(3): 336-43.
[http://dx.doi.org/10.1002/(SICI)1097-4598(200003)23:3<336::AID-MUS4>3.0.CO;2-L]   [PMID: 10679709]

[8]   Wijesekera LC, Leigh PN. Amyotrophic lateral sclerosis. Orphanet J Rare Dis 2009; 4: 3.
[http://dx.doi.org/10.1186/1750-1172-4-3] [PMID: 19192301]

[9]   Leigh P N, Abrahams S, Al-Chalabi A, *et al.* King's MND Care and Research Team (2003).. The management of motor neurone disease. Journal of neurology, neurosurgery, and psychiatry 2003; 74(Suppl 4): iv32-47.

[10]   Gordon PH. Amyotrophic lateral sclerosis: pathophysiology, diagnosis and management. CNS Drugs 2011; 25(1): 1-15.
[http://dx.doi.org/10.2165/11586000-000000000-00000] [PMID: 21128691]

[11]   Cellura E, Spataro R, Taiello AC, La Bella V. Factors affecting the diagnostic delay in amyotrophic lateral sclerosis. Clin Neurol Neurosurg 2012; 114(6): 550-4.
[http://dx.doi.org/10.1016/j.clineuro.2011.11.026] [PMID: 22169158]

[12]   Alsultan AA, Waller R, Heath PR, Kirby J. The genetics of amyotrophic lateral sclerosis: current insights. Degener Neurol Neuromuscul Dis 2016; 6: 49-64.
[http://dx.doi.org/10.2147/DNND.S84956] [PMID: 30050368]

[13]   Brooks BR. El Escorial World Federation of Neurology criteria for the diagnosis of amyotrophic lateral sclerosis. Subcommittee on Motor Neuron Diseases/Amyotrophic Lateral Sclerosis of the World Federation of Neurology Research Group on Neuromuscular Diseases and the El Escorial "Clinical limits of amyotrophic lateral sclerosis" workshop contributors. J Neurol Sci 1994; 124 (Suppl.): 96-107.
[http://dx.doi.org/10.1016/0022-510X(94)90191-0] [PMID: 7807156]

[14]   Brooks BR, Miller RG, Swash M, Munsat TL. World Federation of Neurology Research Group on Motor Neuron Diseases. El Escorial revisited: revised criteria for the diagnosis of amyotrophic lateral sclerosis. Amyotroph Lateral Scler Other Motor Neuron Disord 2000; 1(5): 293-9.
[http://dx.doi.org/10.1080/146608200300079536] [PMID: 11464847]

[15]   Kiernan MC, Vucic S, Cheah BC, *et al.* Amyotrophic lateral sclerosis. Lancet 2011; 377(9769): 942-55.
[http://dx.doi.org/10.1016/S0140-6736(10)61156-7] [PMID: 21296405]

[16]   Conradi S, Ronnevi LO, Norris FH. Motor neuron disease and toxic metals. Adv Neurol 1982; 36: 201-31.
[PMID: 6817611]

[17]   Roth G. Fasciculations and their F-response. Localisation of their axonal origin. J Neurol Sci 1984; 63(3): 299-306.
[http://dx.doi.org/10.1016/0022-510X(84)90152-7] [PMID: 6327921]

[18]   Wettstein A. The origin of fasciculations in motoneuron disease. Ann Neurol 1979; 5(3): 295-300.
[http://dx.doi.org/10.1002/ana.410050312] [PMID: 443761]

[19]   Hogden A, Foley G, Henderson RD, James N, Aoun SM. Amyotrophic lateral sclerosis: improving care with a multidisciplinary approach. J Multidiscip Healthc 2017; 10: 205-15.
[http://dx.doi.org/10.2147/JMDH.S134992] [PMID: 28579792]

[20]   Miller RG, Mitchell JD, Lyon M, Moore DH. Riluzole for amyotrophic lateral sclerosis (ALS)/motor neuron disease (MND). Cochrane Database Syst Rev 2002; (2): CD001447.
[http://dx.doi.org/10.1002/14651858.CD001447] [PMID: 12076411]

[21]   Volk AE, Weishaupt JH, Andersen PM, Ludolph AC, Kubisch C. Current knowledge and recent insights into the genetic basis of amyotrophic lateral sclerosis. Med Genetik 2018; 30(2): 252-8.
[http://dx.doi.org/10.1007/s11825-018-0185-3] [PMID: 30220791]

[22]   Boylan K. Familial Amyotrophic Lateral Sclerosis. Neurol Clin 2015; 33(4): 807-30.
[http://dx.doi.org/10.1016/j.ncl.2015.07.001] [PMID: 26515623]

[23]   Chen S, Sayana P, Zhang X, Le W. Genetics of amyotrophic lateral sclerosis: an update. Mol Neurodegener 2013; 8: 28.
[http://dx.doi.org/10.1186/1750-1326-8-28] [PMID: 23941283]

[24]   Mejzini R, Flynn LL, Pitout IL, Fletcher S, Wilton SD, Akkari PA. ALS Genetics, Mechanisms, and Therapeutics: Where Are We Now? Front Neurosci 2019; 13: 1310.
[http://dx.doi.org/10.3389/fnins.2019.01310] [PMID: 31866818]

[25]   Hübers A, Just W, Rosenbohm A, *et al.* De novo FUS mutations are the most frequent genetic cause in early-onset German ALS patients. Neurobiol Aging 2015; 36(11): 3117.e1-6.
[http://dx.doi.org/10.1016/j.neurobiolaging.2015.08.005] [PMID: 26362943]

[26]   Hayashi Y, Homma K, Ichijo H. SOD1 in neurotoxicity and its controversial roles in SOD1 mutation-negative ALS. Adv Biol Regul 2016; 60: 95-104.
[http://dx.doi.org/10.1016/j.jbior.2015.10.006] [PMID: 26563614]

[27]   Fujisawa T, Homma K, Yamaguchi N, *et al.* A novel monoclonal antibody reveals a conformational alteration shared by amyotrophic lateral sclerosis-linked SOD1 mutants. Ann Neurol 2012; 72(5): 739-49.
[http://dx.doi.org/10.1002/ana.23668] [PMID: 23280792]

[28]   Jiang J, Ravits J. Pathogenic Mechanisms and Therapy Development for C9orf72 Amyotrophic Lateral Sclerosis/Frontotemporal Dementia. Neurotherapeutics 2019; 16(4): 1115-32.
[http://dx.doi.org/10.1007/s13311-019-00797-2] [PMID: 31667754]

[29]   Rosen DR, Siddique T, Patterson D, *et al.* Mutations in Cu/Zn superoxide dismutase gene are associated with familial amyotrophic lateral sclerosis. Nature 1993; 362(6415): 59-62.
[http://dx.doi.org/10.1038/362059a0] [PMID: 8446170]

[30]   Barber SC, Shaw PJ. Oxidative stress in ALS: key role in motor neuron injury and therapeutic target. Free Radic Biol Med 2010; 48(5): 629-41.
[http://dx.doi.org/10.1016/j.freeradbiomed.2009.11.018] [PMID: 19969067]

[31]   Reaume AG, Elliott JL, Hoffman EK, *et al.* Motor neurons in Cu/Zn superoxide dismutase-deficient mice develop normally but exhibit enhanced cell death after axonal injury. Nat Genet 1996; 13(1): 43-7.
[http://dx.doi.org/10.1038/ng0596-43] [PMID: 8673102]

[32]   Wang P, Deng J, Dong J, *et al.* TDP-43 induces mitochondrial damage and activates the mitochondrial unfolded protein response. PLoS Genet 2019; 15(5): e1007947.
[http://dx.doi.org/10.1371/journal.pgen.1007947] [PMID: 31100073]

[33]   Robinson MB. The family of sodium-dependent glutamate transporters: a focus on the GLT-1/EAAT2 subtype. Neurochem Int 1998; 33(6): 479-91.
[http://dx.doi.org/10.1016/S0197-0186(98)00055-2] [PMID: 10098717]

[34]   Kong Q, Carothers S, Chang Y, Glenn Lin CL. The importance of preclinical trial timing - a potential reason for the disconnect between mouse studies and human clinical trials in ALS. CNS Neurosci Ther 2012; 18(9): 791-3.
[http://dx.doi.org/10.1111/j.1755-5949.2012.00358.x] [PMID: 22712693]

[35]   Shaw PJ, Forrest V, Ince PG, Richardson JP, Wastell HJ. CSF and plasma amino acid levels in motor neuron disease: elevation of CSF glutamate in a subset of patients. Neurodegeneration : a journal for neurodegenerative disorders, neuroprotection, and neuroregeneration 1995; 4(2): 209-16.

[36]   Lee SG, Su ZZ, Emdad L, *et al.* Mechanism of ceftriaxone induction of excitatory amino acid transporter-2 expression and glutamate uptake in primary human astrocytes. J Biol Chem 2008; 283(19): 13116-23.
[http://dx.doi.org/10.1074/jbc.M707697200] [PMID: 18326497]

[37]   Bensimon G, Lacomblez L, Meininger V. ALS/Riluzole Study Group. A controlled trial of riluzole in amyotrophic lateral sclerosis. N Engl J Med 1994; 330(9): 585-91.
[http://dx.doi.org/10.1056/NEJM199403033300901] [PMID: 8302340]

[38]   Blokhuis AM, Groen EJ, Koppers M, van den Berg LH, Pasterkamp RJ. Protein aggregation in amyotrophic lateral sclerosis. Acta Neuropathol 2013; 125(6): 777-94.
[http://dx.doi.org/10.1007/s00401-013-1125-6] [PMID: 23673820]

[39]   Ryan VH, Dignon GL, Zerze GH, *et al.* Mechanistic View of hnRNPA2 Low-Complexity Domain Structure, Interactions, and Phase Separation Altered by Mutation and Arginine Methylation. Mol Cell 2018; 69(3): 465-479.e7.
[http://dx.doi.org/10.1016/j.molcel.2017.12.022] [PMID: 29358076]

[40]   Kalmar B, Novoselov S, Gray A, Cheetham ME, Margulis B, Greensmith L. Late stage treatment with arimoclomol delays disease progression and prevents protein aggregation in the SOD1 mouse model of ALS. J Neurochem 2008; 107(2): 339-50.
[http://dx.doi.org/10.1111/j.1471-4159.2008.05595.x] [PMID: 18673445]

[41]   García ML, Fernández A, Solas MT. Mitochondria, motor neurons and aging. J Neurol Sci 2013; 330(1-2): 18-26.
[http://dx.doi.org/10.1016/j.jns.2013.03.019] [PMID: 23628465]

[42]   Borthwick GM, Johnson MA, Ince PG, Shaw PJ, Turnbull DM. Mitochondrial enzyme activity in amyotrophic lateral sclerosis: implications for the role of mitochondria in neuronal cell death. Ann Neurol 1999; 46(5): 787-90.
[http://dx.doi.org/10.1002/1531-8249(199911)46:5<787::AID-ANA17>3.0.CO;2-8] [PMID: 10553999]

[43]   Groeneveld GJ, Veldink JH, van der Tweel I, *et al.* A randomized sequential trial of creatine in amyotrophic lateral sclerosis. Ann Neurol 2003; 53(4): 437-45.
[http://dx.doi.org/10.1002/ana.10554] [PMID: 12666111]

[44]   Shefner JM, Cudkowicz ME, Schoenfeld D, *et al.* NEALS Consortium. A clinical trial of creatine in ALS. Neurology 2004; 63(9): 1656-61.

[http://dx.doi.org/10.1212/01.WNL.0000142992.81995.F0] [PMID: 15534251]

[45]   Smith R, Pioro E, Myers K, *et al.* Enhanced Bulbar Function in Amyotrophic Lateral Sclerosis: The Nuedexta Treatment Trial. Neurotherapeutics 2017; 14(3): 762-72.
[http://dx.doi.org/10.1007/s13311-016-0508-5] [PMID: 28070747]

[46]   He BP, Strong MJ. Motor neuronal death in sporadic amyotrophic lateral sclerosis (ALS) is not apoptotic. A comparative study of ALS and chronic aluminium chloride neurotoxicity in New Zealand white rabbits. Neuropathol Appl Neurobiol 2000; 26(2): 150-60.
[http://dx.doi.org/10.1046/j.1365-2990.2000.026002150.x] [PMID: 10840278]

[47]   Sathasivam S, Ince PG, Shaw PJ. Apoptosis in amyotrophic lateral sclerosis: a review of the evidence. Neuropathol Appl Neurobiol 2001; 27(4): 257-74.
[http://dx.doi.org/10.1046/j.0305-1846.2001.00332.x] [PMID: 11532157]

[48]   Miller R, Bradley W, Cudkowicz M, *et al.* TCH346 Study Group. Phase II/III randomized trial of TCH346 in patients with ALS. Neurology 2007; 69(8): 776-84.
[http://dx.doi.org/10.1212/01.wnl.0000269676.07319.09] [PMID: 17709710]

[49]   Cudkowicz ME, Andres PL, Macdonald SA, *et al.* Northeast ALS and National VA ALS Research Consortiums. Phase 2 study of sodium phenylbutyrate in ALS. Amyotroph Lateral Scler 2009; 10(2): 99-106.
[http://dx.doi.org/10.1080/17482960802320487] [PMID: 18688762]

[50]   Morrison KE, Dhariwal S, Hornabrook R, *et al.* UKMND-LiCALS Study Group. Lithium in patients with amyotrophic lateral sclerosis (LiCALS): a phase 3 multicentre, randomised, double-blind, placebo-controlled trial. Lancet Neurol 2013; 12(4): 339-45.
[http://dx.doi.org/10.1016/S1474-4422(13)70037-1] [PMID: 23453347]

[51]   Hooten KG, Beers DR, Zhao W, Appel SH. Protective and Toxic Neuroinflammation in Amyotrophic Lateral Sclerosis. Neurotherapeutics 2015; 12(2): 364-75.
[http://dx.doi.org/10.1007/s13311-014-0329-3] [PMID: 25567201]

[52]   Aggarwal S, Cudkowicz M. ALS drug development: reflections from the past and a way forward. Neurotherapeutics 2008; 5(4): 516-27.
[http://dx.doi.org/10.1016/j.nurt.2008.08.002] [PMID: 19019302]

[53]   Bachiller S, Jiménez-Ferrer I, Paulus A, *et al.* Microglia in Neurological Diseases: A Road Map to Brain-Disease Dependent-Inflammatory Response. Front Cell Neurosci 2018; 12: 488.
[http://dx.doi.org/10.3389/fncel.2018.00488] [PMID: 30618635]

[54]   Appel S H, Zhao W, Beers D R, Henkel J S. The microglial-motoneuron dialogue in ALS. Acta myologica : myopathies and cardiomyopathies : official journal of the Mediterranean Society of Myology 2011; 30(1): 4-8.

[55]   Ilieva H, Polymenidou M, Cleveland DW. Non-cell autonomous toxicity in neurodegenerative disorders: ALS and beyond. J Cell Biol 2009; 187(6): 761-72.
[http://dx.doi.org/10.1083/jcb.200908164] [PMID: 19951898]

[56]   Beers DR, Henkel JS, Xiao Q, *et al.* Wild-type microglia extend survival in PU.1 knockout mice with familial amyotrophic lateral sclerosis. Proc Natl Acad Sci USA 2006; 103(43): 16021-6.
[http://dx.doi.org/10.1073/pnas.0607423103] [PMID: 17043238]

[57]   Foust K D, Salazar D L, Likhite S, *et al.* Therapeutic AAV9-mediated suppression of mutant SOD1 slows disease progression and extends survival in models of inherited ALS. Molecular therapy : the journal of the American Society of Gene Therapy 2013; 21(12): 2148-59.

[58]   Li L, Liu J, She H. Targeting Macrophage for the Treatment of Amyotrophic Lateral Sclerosis. CNS Neurol Disord Drug Targets 2019; 18(5): 366-71.
[http://dx.doi.org/10.2174/1871527318666190409103831] [PMID: 30963986]

[59]   Ferreira R, Santos T, Gonçalves J, *et al.* Histamine modulates microglia function. J Neuroinflammation 2012; 9: 90.

[http://dx.doi.org/10.1186/1742-2094-9-90] [PMID: 22569158]

[60]   Nussbacher JK, Tabet R, Yeo GW, Lagier-Tourenne C. Disruption of RNA Metabolism in Neurological Diseases and Emerging Therapeutic Interventions. Neuron 2019; 102(2): 294-320.
[http://dx.doi.org/10.1016/j.neuron.2019.03.014] [PMID: 30998900]

[61]   Lagier-Tourenne C, Polymenidou M, Cleveland DW. TDP-43 and FUS/TLS: emerging roles in RNA processing and neurodegeneration. Hum Mol Genet 2010; 19(R1): R46-64.
[http://dx.doi.org/10.1093/hmg/ddq137] [PMID: 20400460]

[62]   Maxwell MM. RNAi applications in therapy development for neurodegenerative disease. Curr Pharm Des 2009; 15(34): 3977-91.
[http://dx.doi.org/10.2174/138161209789649295] [PMID: 19751205]

[63]   Lepore E, Casola I, Dobrowolny G, Musarò A. Neuromuscular Junction as an Entity of Nerve-Muscle Communication. Cells 2019; 8(8): 906.
[http://dx.doi.org/10.3390/cells8080906] [PMID: 31426366]

[64]   Cruz MP. Edaravone (Radicava): A Novel Neuroprotective Agent for the Treatment of Amyotrophic Lateral Sclerosis. P &T : a peer-reviewed journal for formulary management 2018; 43(1): 25-8.

[65]   Tosolini AP, Sleigh JN. Motor Neuron Gene Therapy: Lessons from Spinal Muscular Atrophy for Amyotrophic Lateral Sclerosis. Front Mol Neurosci 2017; 10: 405.
[http://dx.doi.org/10.3389/fnmol.2017.00405] [PMID: 29270111]

[66]   Blessing D, Déglon N. Adeno-associated virus and lentivirus vectors: a refined toolkit for the central nervous system. Curr Opin Virol 2016; 21: 61-6.
[http://dx.doi.org/10.1016/j.coviro.2016.08.004] [PMID: 27559630]

[67]   Nizzardo M, Taiana M, Rizzo F, *et al.* Synaptotagmin 13 is neuroprotective across motor neuron diseases. Acta Neuropathol 2020; 139(5): 837-53.
[http://dx.doi.org/10.1007/s00401-020-02133-x] [PMID: 32065260]

CHAPTER 9

Impact of Nano-Formulations of Natural Compounds in the Management of Neuro degenerative Diseases

Hemraj Heer[1], Vishav Prabhjot Kaur[1], Tania Bajaj[1], Arti Singh[2], Priyanka Bajaj[3] and Charan Singh[1,*]

[1] *Department of Pharmaceutics, ISF College of Pharmacy, Moga, Punjab-142001, India*

[2] *Department of Pharmacology, ISF College of Pharmacy, Moga, Punjab-142001 Affiliated to IK Gujral Punjab Technical University, Jalandhar, Punjab-144603, India*

[3] *Institiute of Microbial Technology, Sector 39A, Chandigarh-160036, India*

Abstract: Neurodegenerative disorders (NDs), such as Alzheimer's disease (AD), Parkinson's disease (PD), and Huntington's disease (HD), are caused by oxidative stress, inflammation, and proteinopathy. These are further characterized by loss of neurons and, consequently, impaired cognitive functions. However, the exact mechanisms of the pathogenesis of these diseases are still unknown. Nowadays, natural compounds like curcumin, quercetin, resveratrol, and piperine, among others, have been explored for the treatment and prevention of neurological disorders. There are various *in vivo* studies and clinical trials conducted for alleviating neurological disorders using natural compounds encapsulated in nanocarrier systems. Nanoparticles such as lipidic, polymeric, quantum dots help to enhance the bioavailability, specificity, and targeted delivery of these compounds in the brain. Various simple and reproducible methods are reported to synthesize the nanoparticles in the literature. In this chapter, we will explore the role of nanotechnology and natural compounds to treat and prevent neurodegenerative disorders.

Keywords: Nanotechnology, Nanoparticles, Natural compounds, Neurodegenerative disorders, Targeted delivery.

INTRODUCTION

Neurodegenerative disorders affect millions of people worldwide every year. The vitality and functionality of the nerve or central nervous system (CNS) are affected partially or completely due to Alzheimer's disease (AD), Parkinson's

---

* **Corresponding author Charan Singh:** Department of Pharmaceutics, ISF College of Pharmacy, Moga, Punjab-142001, India; Tel: +91 9817067168; E-mail: c.singhniper09@gmail.com

Sachchida Nand Rai (Ed.)

disease (PD), Huntington's disease (HD), Multiple Sclerosis (MS), and Amyotrophic Lateral Sclerosis (ALS). At the same time, it offers a great challenge owing to poor prognosis *vis-a-vis* the rest of the body part [1]. The measure cause of neurodegeneration is the loss of myelin sheath, protein degradation, mitochondrial dysfunction, accumulation of mutated proteins, family background along with some environmental factors. Furthermore, aging is another important mechanism that is reported in various brain disorders [2]. AD is considered the most common form of dementia in which cognitive functions alter. As per Alzheimer's Disease (AD) Report 2019, around 6 million Americans are living with Alzheimer's dementia [3]. Extracellular accumulation of β-amyloid (Aβ) peptide and deposition of tau-protein cause neuroinflammation in the brain and, consequently, synaptic impairment and neuronal loss [4]. PD is considered the second most prominent ND. This occurs due to the development of bradykinesia and tremors of cardinal motor functions in the *substantia nigra* [5]. The major hallmark of PD is reducing uptake of dopamine by dopaminergic neurons due to the low level of dopamine transporters (DATs). Additionally, the deposition of α-synuclein in Lewy bodies is another cause of PD [6]; however, their actual mechanism to produce Parkinson's is not fully understood yet [7]. Huntington's disease (HD) is another autosomal ND that usually develops due to mutation in genes located on chromosome 4. This genetic disorder is characterized by repeated expression of CAG (cytosine, adenine, and guanine) tri-nucleotides. The HD progresses due to the accumulation of Huntington mutated protein as a result of the repeated unit of CAG [8]. Behavior abnormality and change gate are the common symptoms in this disease [9].

Some of the common factors involved in neurodegeneration are the level of glutamate, free radical generation, the concentration of reactive nitrogen species, proteinopathy, nuclear pore anomalies, and inflammation, and calcium ion [10]. Glutamate is an excitatory neurotransmitter in the brain, and its overexcitation activates N-methyl-D-aspartate (NMDA) and α-amino-3-hydroxy-5-me-hyl-4-isoxazole propionic acid (AMPA), eventually leading to the apoptotic destruction of neurons. Additionally, proteinopathy is responsible for the accumulation of misfolded proteins such as α-synuclein and Aβ in PD and AD, respectively [11]. Elevated levels of these proteins are reported in the clinical conditions of neurodegeneration. Moreover, free radical and reactive oxygen species holds the lion's share. These are the radicals with lone pairs of electrons that can trigger free radical cascade and neuron damage by disturbing its biochemistry [12]. The influx and efflux of proteins like nucleoporins (Nups) and Ran GTPase-activating protein (RanGAP) across the nuclear barrier play a key role in the normal functioning of the neuron [13]. Hence, impairment of influx and efflux proteins causes damage to the nuclear physiology, which subsequently induces neurodegeneration. In recent times, neuroinflammation has been another dimension in the aetiology of various NDs that has been under investigation. Scientists have found the significant role of inflammatory mediators and calcium in the progress of neuronal loss [14]. Calcium is an important mediator involved in neuronal physiology and integrity. Hence, abnormal calcium transport and barrier function can lead to neuroinflammation, apoptosis, and loss of neurons [15].

NEUROPROTECTIVE POTENTIAL OF NATURAL PRODUCTS

Tremendous efforts have been made to introduce new CNS drugs for effective treatment as available drugs are mainly used for symptomatic relief. However, designing permanent therapeutics or preventive medicines for curing and/or preventing neurodegeneration is still a great challenge. Various synthetic and semisynthetic therapeutic agents have remarkable therapeutic effects in disease management, but severe side effects limit their use [16]. Nowadays, natural products have gained attention as new promising therapeutic agents due to their neuroprotective, antioxidant, and anti-inflammatory activities [17]. Ample reports suggest the role of inflammatory markers and reactive species for neuronal pathology through biological pathways [13]. Therefore, inhibition of these pathways using natural compounds might play a key role in the management of NDs [18]. Of late, fighting NDs using herbal drugs has become a thrust area. After The 'Green' movement in Western society, the majority of health concerned people in these countries utilized phytomedicine for primary healthcare as various plants-based bioactives were used in traditional medicine to cure neurodegenerative diseases. The natural product contains diverse phytoconstituents in the form of fatty acids, sterols, alkaloids, flavonoids, glycosides, saponins, tannins, terpenes, *etc.* These possess strong therapeutic potential to treat different types of diseases [19]. There are more than 120 medicinal plants and their natural compounds that are used to cure neurodegenerative disease, and they show promising neuropharmacological activity [20]. A few of them are Curcumin, *Ginkgo biloba, Panax ginseng, Bacopa monniera*, Withania somnifera, Polyphenols (Epigallocatechin-3-galate (EGCG) Resveratrol), Flavonoids Quercetin Terpenoids, and Saponins [21 - 34].

APPROACHES FOR CNS TARGETED NATURAL COMPOUNDS DELIVERY

In this modern era, even though sufficient knowledge and new inventions in the field of medicine and drug delivery techniques are available, still the whole world is fighting with the most complex neurodegenerative diseases such as Alzheimer's, Parkinson's, Huntington's disease, Multiple Sclerosis, and Amyotrophic Lateral Sclerosis, *etc.* The limited clinical interventions in neurodegenerative diseases due to insufficient drug accessibility in the brain slow down disease prognosis [35]. The major hurdle for developing new strategies of CNS treatment is the protective blood-brain barrier (BBB), which comprises tight junctions connected through polarized endothelial cells with fine blood capillaries and enormous transporters, which further limits the passage of drugs to the CNS region [36]. Moreover, non-targeted drug deliveries, as well as unwanted adverse effects of diagnostic reagents on neuronal cells, emphasize designing and developing a novel theranostic approach for neurological disorders. Therefore, the different approaches are being exploited by various researchers not only to improve the efficacy of the treatment methodologies but also provid the medical experts with the tools that can provide an early and timely diagnosis of the medical condition [37]. Novel strategies that have been devised and put to practical use include BBB disruption, receptor-mediated brain targeting, bypassing BBB *via a* nasal route through the olfactory epithelium, and usage of nano-size delivery systems. In the next section of the book chapter, various nanotechnological strategies from liposomes to advanced nanocarriers are discussed in detail.

NANO-BASED DELIVERY SYSTEMS FOR NEURODEGENERATIVE DISORDERS

Nanotechnology introduces a variety of new drug delivery systems capable of overcoming various limitations of neuropsychiatric drugs. This multidisciplinary novel research area includes physics, chemistry, and biology concerned with the manipulation in size ranging from 10- 100 nm at the atomic and molecular level [38]. The high stability, large payload, and flexible design make nanomaterials a powerful drug delivery carrier to cross the BBB [39]. Nanomaterials have higher cellular uptake compare to traditional drugs, which allows nanomaterials to target a wide range of both cellular and intracellular targets. Furthermore, optimum *in vivo* pharmacokinetic profile, target-oriented delivery of materials enhance the safety, sensitivity, and personalization of nano-carriers for various applications like diagnosis, imaging, and to treat various neurological disorders [40]. A different class of nano delivery systems is meant for different applications based on the material composition, size range, *in vivo* profile like lipid-based nanoparticles, carbon-based nanoparticles, and magnetic nanoparticles are developed to diagnose and treat neuropsychiatric disease. Some modifications in nano cargos improve circulation time and control the drug release, such as polyethylene glycol (PEG) and poly (lactic-coglycolic acid) (PLGA). Nanomedicine is considered a promising candidate to release a sufficient amount of drug in the brain by crossing BBB. Similarly, other exclusive features like more resolve image, nano-magnetic, and photothermal effects make them an ideal material for diagnosis. Moreover, nano-sized delivery system like nanobubbles has theranostic feature a platform to combine diagnosis and drug delivery simultaneously. The various advancements in the nanotechnology field provide opportunities to treat and diagnose of neuropsychiatric disease more efficiently [41].

Numerous nano-size-based drug delivery systems for AD and PD have been developed however, this number is low for the treatment of HD. The various nano-sized drug delivery systems are discussed in more detail in the later section of this chapter. A more common form of NDs such as AD or PD is associated with symptoms like damage to neurons with eventual limb paralysis, loss of speech, swallowing and breathing functions, and eventually death [42]. Nano-drug delivery systems significantly contribute to the design of different therapeutics approaches by modifying the pharmacological properties of drugs and related therapeutic responses, which allow them to achieve a better pharmacokinetic profile in the presence of other factors. A wide range of nano-drug delivery systems has been developed, ranging from nanofibers to polymeric nanomicelles, which differ from each other in terms of geometries, configurations, and surfaces ranging from [43]. Fig. (1) illustrates various types of nanocarriers systems for delivering natural compounds for the treatment of NDs.

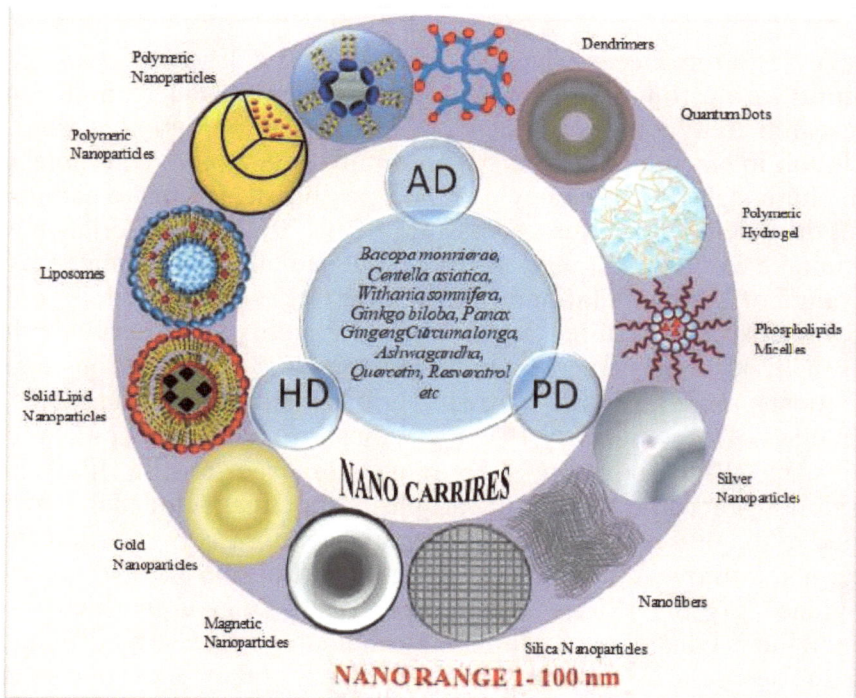

Fig. (1). Nanocarriers used for delivering natural compounds.

LIPIDIC NANOMATERIALS

Liposomes

Liposomes are the vesicular drug delivery system; comprises of many phospholipid layers that have the potential to encapsulate and transfer a variety of cargo [44]. These were discovered in 1961 by Alec D Bangham. The surface decoration with various agents has been carried out in improving its targetability (surface tethering of RGD, KDEL peptides) and/or improving the circulation time (surface PEGylation of the liposomes). Moreover, the minuscule size in the range of 100 nm makes it an eligible candidate to smuggle the cargo across the BBB [45].

Solid Lipid Nanoparticles and Nanostructured Lipid Carrier Systems

Introduced in 1991, solid lipid nanoparticles (SLNs) successfully emerged as an alternate drug delivery system to emulsions, liposomes, suspensions, *etc.* These contain lipids that are solid at room temperature. Typically, comprise of a solid lipid(s), surfactant(s), co-surfactant (optional), and active principles (generally drugs or bioactives). These systems can be tailored as per the requirement employing surface decoration with myriad types of ligands for achieving theranostic attributes [46]. However, SLNs possess issues such as low drug payload, long-term stability, and drug leakage. Hence, with a slight modification using the addition of liquid lipids, nanostructured lipid carriers (NLCs) adds some stellar attributes to their credit. The liquid lipid enhances the solubility, permeability as well as carrying capacity of the drugs. This ultimately results in enhanced bioavailability and reduction in the adverse effects associated thereafter [47]. The nanolipids brain uptake mechanisms vary with the change in system composition and surface modifications. Besides providing a hydrophobic lipidic composition, in some cases, the employed carrier may additionally limit the binding to P-gp and avoids efflux, thereby leading to higher brain drug concentration. Preferential BBB uptake can also be hypothesized owing to the surface modifications with positive charge inducers. Numerous studies have proved that positively charged lipidic nanoparticles are preferentially taken up by the brain endothelial cells [48].

Carbon Nanomaterials

Carbon-based nano-engineering has been a global trend in drug delivery research for a while now. Carbon, because of its unique allotropic attributes, manifests itself into various physical forms, each with special qualities and potential that can be harnessed to come up with a novel nanowagon that is safe as well as effective [49]. The superfamily of carbon comprises the diverse genre of carbon nanosystems. These include carbon nanotubes (CNT), fullerenes, graphenes, and nanodiamonds. In context with the brain targeting for NDs, these carbon nanosystems have shown exceptional capability in terms of targetability, suitability, and carrying the potential for a diverse genre of cargo [50]. Here we discussed, the two most widely used carriers belonging to carbon materials.

Carbon Nanotubes

Carbon nanotubes (CNTs) are tubular nanoscaled systems that have been used to target at the cellular level [51]. The hollow structure aids in the easy carrying capability. The unique aspect ratio, enormous surface area, and optical properties make this allotrope unique. After certain modifications in the chemistry of CNTs, *via* certain chemical processes, functionalized CNTs could be prepared [52]. However, Pristine CNTs have an issue of toxicity and bioaggregation in the biological milieu. Therefore, decoration of the surface using some ligands could be an approach to reduce their toxicity. Owing to their unique aspect ratio and minuscule size can easily traverse the BBB and have been experimentally proved to be quite instrumental in NDs. Beside, graphenes and fullerenes are also being explored in the area of brain delivery nowadays [53]

Nanodiamonds

A carbon allotrope, nanodiamonds have been explored for their theranostic applications, a high degree of fluorescence capability, and the ability to resolute to a very minute level. Owing to these properties, these are perfect candidates for neuronal theranostic applications [54]. Moreover, the non-toxicity of these carbon nanosystems and their resistance to *in-vivo* degradation for an extended period makes them a perfect biomedical imaging tool using magnetic resonance [55].

Polymeric Systems

Polymeric systems are the most versatile and widely used nanocarriers due to their enormous advantages such as biocompatible, biodegradable nature, easy surface functionalization, ability to encapsulate hydrophobic and hydrophilic drugs, *etc.* Nanosystems of the various architectural forms and compositions have been explored for neurodegeneration [56]. The polymers used can be natural, synthetic, and/or semi-synthetic in origin. These systems have been used to carry a variety of cargo across the BBB [57]. To overcome the challenges of biofate and therapeutic efficacy, biodegradable versions of polymeric nanosystems have been used. Additionally, scientists have explored different types of polysaccharides and polypeptides for the said purpose. Experimental strides have unearthed the potential of such polymeric nanosystems to be explored for myriad theranostic applications in the milieu of neurodegeneration [58]. Moreover, these systems possess good CNS biocompatibility and can be appended with several brain targeting ligands. Based on the type of ligand used, a variety of transcellular transport processes across the BBB can be distinguished. The most common pathway is diffusion, which involves the transfer of small hydrophobic molecules into the brain [59].

Dendrimers

Dendrimers are hyperbranched, tree-like nanosystems with multiple layers. These layers are referred to as the generations of the dendrimers. These are promising systems owing to their high targeting potential, low toxicity, and high specificity. Additionally, the functionalization of these systems grants them various theranostic attributes [60]. Dendrimers have shown great potential not only for diagnostic purposes but in the treatment of several CNS diseases also. With the application of these nanocarriers, it is possible to target ultra-low concentrations of protein biomarkers (Aβ, α-syn, and Huntington) and the cells (activated microglia and astrocytes) involved in the NDs [61].

Metal-organic Frameworks (MOFs)

MOFs are nonporous hybrid nanomaterials comprising 3-Dimensional arrays of the various metals tethered to an organic linker [62]. MOFs have been functionalised with different types of diagnostic and imaging agents that make them promising candidates to be explored for CNS disorders. In the drug delivery milieu, MOFs have shown an exceptional capability with high drug loading efficacy and the ability to carry across the barriers. Some scientific reports suggest their application in the area of neurological theranostic [63]. Their capabilities to target the truncated proteins present in the pathogenesis of various NDs, making them promising candidates for nanoneuromedicine [64].

Semiconducting Nanoparticles And Quantum Dots

These fluorescent nanocrystals constitute a unique genre of nanoscaled systems. Quantum dots (QDs) have the size of the order of Bohr's radius (0.529 Å). Owing to such minuscule size, some stellar properties are exhibited by these nanosystems in terms of their fluorescence, *etc.* QDs have been explored for imaging, biosensors, and molecular probes application [65]. Their composition makes them resistant to any damage caused by the high energy radiations during the fluorescence, unlike the fluorescent dyes, which exhibit the bleaching effect. Bioconjugated QDs have been extrapolated as the vectors for the excellent fluorescent probe activity across the BBB [66].

Phospholipid Complexes

These are also known as pharmacosomes or phytosomes based on the nature of the drug encapsulated in the phospholipid matrix. Phospholipids are an integral part of biomembrane and act as a carrier system for delivering hydrophobic as well as hydrophilic drugs [67]. The biocompatible and biodegradable phospholipid complex has been another approach to deliver therapeutics across BBB. Incorporation of the active principles into the phospholipid layers tends to increase the solubility as well as the permeability of the neurotherapeutics. Various studies have suggested marked improvement in the biopharmaceutical attributes of the drugs for brain targeting *via* complex formation [68].

METHODS OF NANOFORMULATIONS PREPARATION

There are two approaches for the synthesis of nanoparticles: top-down and bottom-up techniques. In the top-down method, there is the formation of small particles *via* dissociation of large particles, while the latter is concerned with the formation of a complex structure through the orientation of smaller particles. Generally, bottom-up techniques are used for the synthesis of conventional nanoparticles. Moreover, the selection of an appropriate technique also depends on the nature of the drug and physicochemical properties of the polymer [69]. Various methods are described in this section.

Emulsion-solvent Evaporation Method

This is the most frequent method and comprises two steps. Firstly, the emulsification of polymer solution in the aqueous phase followed by evaporation of the solvent under optimized conditions and thereafter, collection of nanoparticles. Briefly, the drug is mixed with the polymer in a suitable organic solvent and subjected to emulsification in an aqueous solution using high shear homogenizer; thereby, the polymer encapsulates drugs in the form of nanoparticles. Then, the solvent is evaporated using rotavapor or slowly increasing temperature with continuous stirring. In this method, various parameters can be controlled to achieve the desired size, such as stirring speed, type and amount of solvent, the viscosity of organic and aqueous phases, temperature, *etc.* This is an economical method used for the preparation of diverse types of emulsions of hydrophobic drugs. Some of the polymers used are poly-lactic acid, poly-lactic glycolic acid, cellulose derivative, poly (E-caprolactone), and poly(h-hydroxybutyrate) [70]. Fig. (**2**) shows the schematics representation of the elusion-solvent evaporation method.

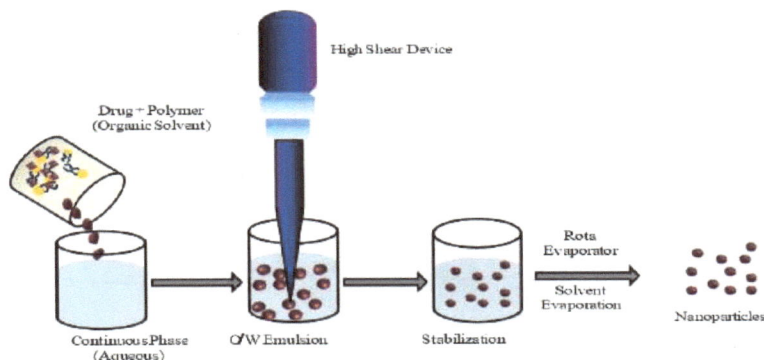

Fig. (2). Emulsion-solvent evaporation method.

Solvent Diffusion Method

The emulsification solvent diffusion or solvent diffusion method is quite similar to the solvent evaporation method as presented in Fig. (**1**). However, this method involves the dissolution of polymer in a partially water-soluble solvent (propylene carbonate). The polymer attains initial thermodynamic equilibrium after saturated with water upon mixing. Thereafter, polymer gets precipitated upon high-speed emulsification and consequently leads to the formation of nanoparticles. Finally, the solvent is removed under constant stirring or a rotatory evaporator. This method has various advantages such as rapid encapsulation, high reproducibility, and easy scalability, simple and economical. However, the major drawback is the leakage of hydrophilic drugs, as well as a large volume of water, which is difficult to eliminate easily [71]. Fig. (**3**) shows the solvent diffusion method.

Salting Out Method

This technique is a modified form of the above-discussed methods. The salting-out method is based upon the principle of separation of water-miscible solvent from aqueous solution *via* a salting-out effect, as displayed in Fig. (**4**). In this process, drug and polymers are dissolved in an organic solvent and then transferred in aqueous media containing salting-out ingredients (electrolytes like $MgCl_2$, $CaCl_2$ or non-electrolytes such as sucrose) and stabilizers such as polyvinylpyrrolidone or hydroxyethylcellulose for emulsification using a high-speed shear device. Thereafter, the emulsion is diluted with a sufficient amount of water to accelerate the diffusion of the organic phase into the aqueous phase, thus inducing the formation of nanoparticles. The obtained nanoparticles are then separated from the salting-out medium through cross-flow filtration assembly. The type of salting-out agent affects the drug loading efficiency. This method is suitable for proteins and thermolabile substances. However, not used for hydrophilic drugs [72]. Fig. (4) represents the salting-out method.

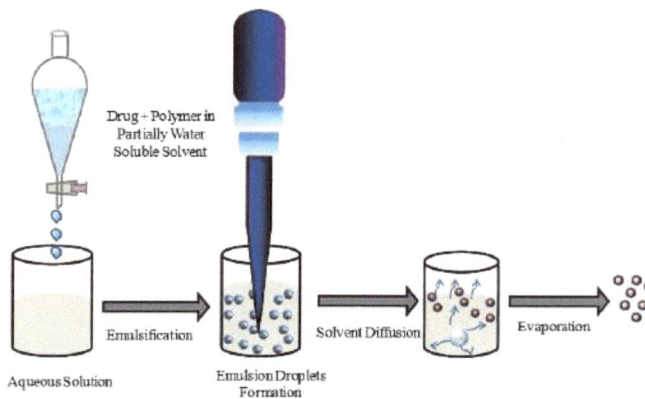

Fig. (3). Solvent-diffusion method.

Supercritical Fluid Technique

The state above the critical temperature (CT) and critical pressure (CP) of any substance is known as its "Supercritical" state, and the substance in this particular phase is known as supercritical fluid (SCF). In a supercritical state, the density of a substance is similar to the same substance in its liquid state. Thus, this property helps to enhance the solubility of poorly aqueous soluble drugs. The freely soluble solutes in SCF are used in this method [73]. The most widely used SCF is $scCO_2$, which dissolves drug or drug-polymer matrix in the high-pressure compression chamber. These are, then, allow passing through a heated nozzle into a low-pressure chamber. The sudden decompression in SCF saturated with solute under nucleation results in nanoparticle formation, as shown in Fig. (5). The major advantage of SCF based nanoparticle formation technique includes preservation of particle properties (morphology and size), eco-friendly, and suitable for mass production. However, agglomeration of nanoparticles is a common problem; can be solved using different chambers for the collection of different sizes [74]. Fig. (**5**) shows a supercritical fluid technique.

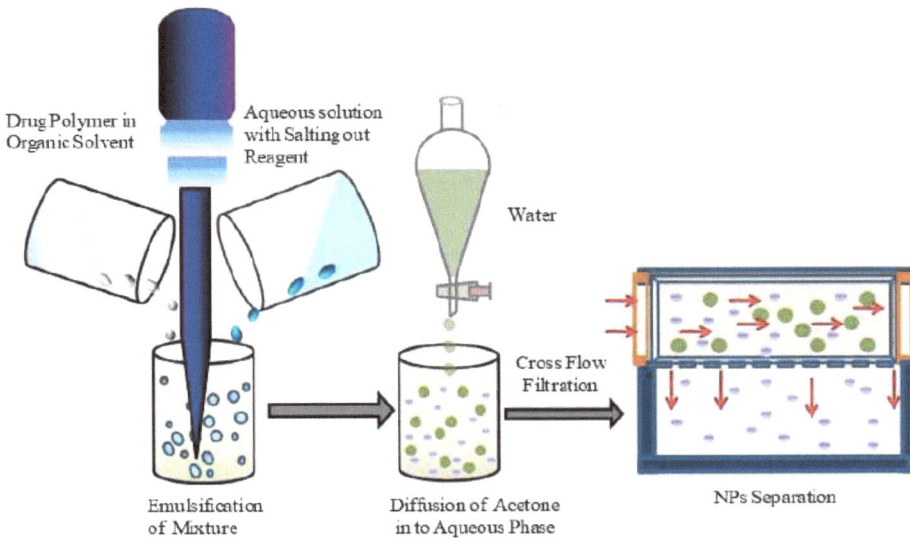

Fig. (4). Salting-out method.

Polymerization Technique

Emulsification or polymerization is considered a rapid technique for nanoparticle synthesis. This process can be two types based on the type of continuous phase involved. When the continuous phase is organic, monomers are dissolved in the organic phase, and interfacial polymerization of monomers is formed around the aqueous phase encapsulating drug material in the form of nanoemulsion and vice versa if the aqueous phase is continuous. The slower polymerization kinetics and dissolution of reactants in water are required for the successful synthesis of nanoparticles. Polymerization kinetics affects the particle size in microemulsion and miniemulsion processes. The former process results in a particle size range of less than 80nm [75]. Fig. (**6**) demonstrates the polymerization technique for the synthesis of nanoparticles. Fig. (**6**) shows polymerization techniques.

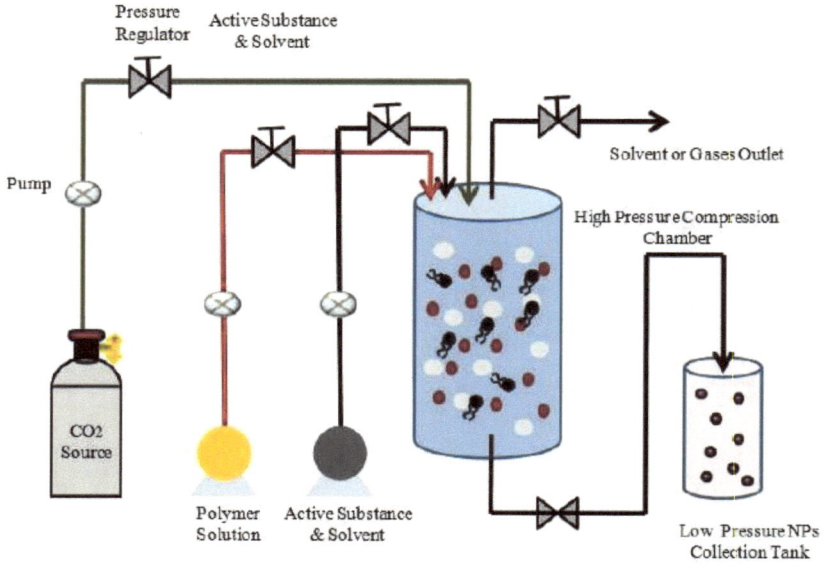

Fig. (5). Supercritical fluid technique.

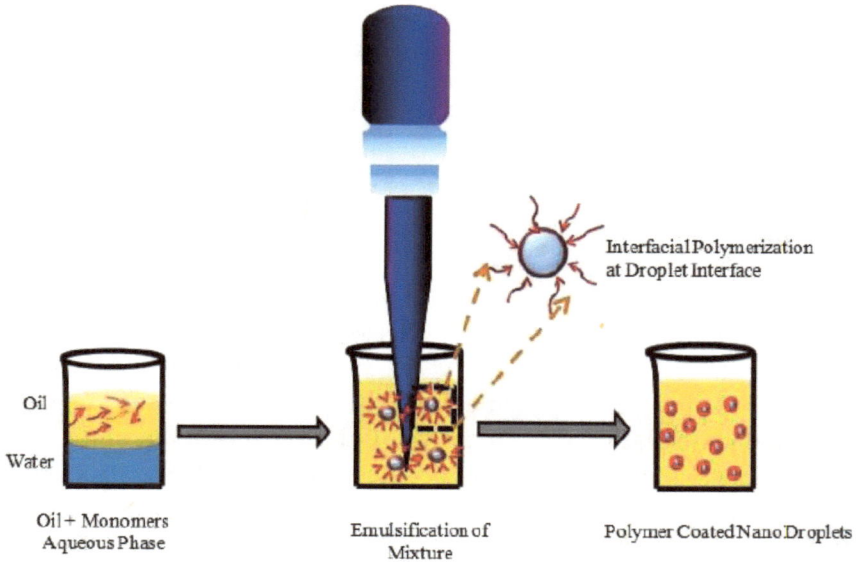

Fig. (6). Polymerization technique.

NATURAL COMPOUNDS NANO-DELIVERY IN NEURO DEGENERATIVE DISEASES

Alzheimer's Disease

In a recent study, quercetin encapsulated polysorbate 80 coated AuPd core-shell nanostructures were prepared to treat AD. The findings of the studies revealed that colocalization of LC3 and lysosome promoted lysosome and autophagosome fusion and their degradation in SH-SY5Y cells. Moreover, the MTT assay illustrated reduced Aβ mediated cytotoxicity and accelerated the clearance of Aβ after treatment. In addition, the transwell assay demonstrated enhanced permeation for BBB permeability and biocompatibility [76]. In another similar study, Pinheiro and co-workers prepared lipid nanoparticles of quercetin. Thereafter, LDH assays highlighted no toxicity in *in-vitro*. Moreover, enhanced permeation was expressed by NLCs across hCMEC/D3 cell monolayers. Simultaneously, Aβ studies demonstrated NLC modified with transferrin has the potential to retard fibril formation [77].

Dongdong *et al.* investigated the effects of PLGA-functionalized quercetin nanoparticles to inhibit and dissemble Aβ fibrils. Strategically, nanoparticles exhibited negligible toxicity on the SH-SY5Y human neuroblastoma cells and inhibited neurotoxicity. Furthermore, behavioral experiments illustrated that *in vivo* inoculation of nanoparticles into APP/PS1 mice ameliorates cognition and memory impairments. In addition, histological investigations showed a null sign of toxicity after the nanoformulations administration [78].

Another study reported the role of quercetin nanoparticles in scopolamine-induced spatial memory deficits, incorporating into the polymeric particulate matrix to get a synergistic effect of polymer and drug. As per results, pretreatment with nanoparticles showed a significant decrease in MDA, AChE levels and an increase in brain catalase and GSH levels relatively equal to the standard group. Furthermore, behavioral, biochemical, and histological investigations established quercetin nanoparticles might be a preventive strategy against the progression of AD [79].

In another study, quercetin was incorporated in Zein nanocapsules and evaluated for various biopharmaceutical attributes after oral administration. Afterthe synthesis and characterization, nanoformulations were given to SAMP8 mice orally, and animals were observed for *in-vivo* attributes. Observations demonstrated improved cognition and memory impairments attributes in the case of nanoparticles in contrary to free quercetin. Furthermore, the biodistribution study illustrated significant levels of quercetin in the brain of mice, quantitatively the post NPQ treatment [80].

As discussed in the introduction part, oxidative stress is among the leading causes of Alzheimer's, which induces an irreversible disruption of synaptic connectivity and death of neurons. In this study, naringin loaded PEG-3000 silica nanoparticles were prepared. The findings demonstrated that naringin pegylated nanoparticles expressed an enhanced protective role against Aβ linked oxidative stress in primary rat neuronal and glial hippocampal cultures [81].

Elham and colleagues reported a comparative study of hesperetin and hesperitin nanoparticles to combat AD progression *via* an antioxidant effect. They induced AD in animals by intracerebroventricular injection of streptozocin (icv-STZ) unilaterally, later treated with hesperetin and hesperitin nanoparticles. They further reported that behavioral analysis depicted improved memory retrieval and recognition memory consolidation. Furthermore, nanoformulations elevated the activity of antioxidant enzymes while declined MDA in the hippocampal area [82].

Lu *et al.* investigated polymeric micelles loaded resveratrol on PC12 cells. The comparative *in-vitro* analysis established that free resveratrol exhibited toxicity to PC12 and did not protect PC12 cells against Aβ induced oxidative stress. On the other hand, the equivalent concentration of nanoparticles yielded better therapeutic results by attenuating oxidative stress; moreover, no signs of cytotoxicity were observed. The authors concluded that the polymeric delivery of resveratrol efficiently treated AD hallmarks in *in-vitro* studies [83].

Resveratrol is supposed to act by removal of β-conformation to reconverse Aβ formation as aggregation leads to a formation of insoluble fibrils. To overcome this, in a study, authors employed a synergistic approach by combining resveratrol with grape seed extract as it aggravates the inhibition effect on Aβ aggregation. After characterization, the invading capacity of the carrier matrix was evaluated using an *in-vitro* model of human BBB [84]. In another investigation, resveratrol loaded polymeric lipid core nanocapsules potentially delivered high concentrations in the brain tissue with remarkable therapeutic efficacy [85].

Another comparative study for functional and non-functional drug carrier matrix was reported by Yang and associates. Here, selenium was added to the resveratrol nanoparticles. *In-vitro* biological studies implied that a synergistic effect was exhibited by Se incorporated nanoparticles. Further, the MTT assay showed no sign of cytotoxicity. Additionally, binding affinity confirmed the interaction between Aβ molecules and nanoparticles. Finally, the author suggested that before clinical application, these should be investigated for *in-vivo* study [86].

In another study, lipid nanostructures are embedded with ferulic acid and investigated for cytotoxic, cell viability, and apoptosis interactions in LAN 5 cell lines. Herein, the cellular invasion capacity of the carrier system was confirmed by light field and fluorescent microscopy. Furthermore, increased vitality was observed for lipidic carriers than free drugs suggesting that nanoparticles act as nano-vectors to deliver drugs efficiently. Moreover, FA-loaded SLNs caused a reduced ROS production in cells hence declined oxidative stress [87].

Curcumin is known to block Aβ peptide by binding a catalyst of neurodegeneration, and preventing agglomeration, but it is challenged due to restriction of crossing BBB. Hence, Cheng *et al.* employed a nanotechnology approach to overcome the mentioned limitation of curcumin. After synthesis, it was evaluated for *in-vivo* behavior in Tg2576 transgenic mice through observing radial arm maze and contextual fear conditioning memory test manners, along with pharmacokinetic attributes. Further distribution studies suggested better biopharmaceutical attributes of nano curcumin. Additionally, positive responses for amyloid plaque density, pharmacokinetics, and Madin–Darby Canine Kidney cell monolayer penetration strengthen the strategy of nano curcumin [88]. In another study, the authors aimed to deliver curcumin polymeric nanoparticles liganded with Tet-1peptide. Functionalization with Tet-1 peptide enhances the neurons targeting and possess retrograde transportation properties. The results demonstrated that curcumin encapsulated-PLGA nanoparticles potentially destroyed amyloid aggregates, exhibited antioxidative property without any signs of cytotoxicity [89].

In a similar study, authors aimed to develop multi-interventional strategies to deliver curcumin efficiently in an *in vitro* neuronal AD cell model to inhibit the caspase-mediated apoptosis along with the evaluation of anti-amyloid and anti-inflammatory effects of the drug. The findings of the studies demonstrated that nano curcumin declined the Aβ levels remarkably, exhibited better antioxidant activity, and increased cell viability contrary to curcumin [90]. Similarly, Na *et al.* employed combination therapy *via* nanoencapsulated curcumin with Aβ generation inhibitor S1 (PQVGHL peptide) conjugated CRT (cyclic CRTIGPSVC) peptide [91]. In another similar study, Fan *et al.* employed different targeting ligand B6 peptides for conjugating curcumin-loaded PLGA-PEG nanoparticles. Morris water study concluded that PLGA-PEG-B6/Cur could tremendously improve the spatial learning and memory capability of APP/PS1 mice, compared with native Curcumin. Additionally, *ex vivo* assays illustrated that PLGA-PEG-B6/Cur could reduce hippocampal β-amyloid formation [92].

Recently, for better therapeutic delivery of curcumin with the aid of selenium fabricated PLGA-Cur nanoparticles were reported. The current study showed that the formed nanocarriers system could decrease the Aβ load in AD mice's brains and efficiently improved memory deficiency. Additionally, Se-PLGA targeted delivery system provides enhanced therapeutic efficacy in AD lesions [93]. Similarly, Yung-Chih *et al.* reported a combination therapy of curcumin with rosmarinic acid PLGA nanoparticles conjugated with 83-14 monoclonal Antibody [94]. In another investigation, curcumin-conjugated nanoliposomes were delivered successfully for better therapeutic effects [95]. Parallel to other investigations, curcumin tagged lipidic nanostructures expressed a high binding affinity for Aβ fibrils while maintaining planar configuration [96].

Another photo element Huperzine A was investigated as a therapeutic agent in AD treatment *via* intranasal delivery. Findings of the studies showed lower toxicity in the 16HBE cell lines compared with drug solution and better cellular uptake in comparison to nontargeted analogs in 16HBE and SH-SY5Y cells, qualitatively and quantitatively. Moreover, *in vitro* studies revealed a longer residence time than nontargeted NPs [97]. In another study, Huperzine A encapsulated lipidic nanocarriers (SLNs, NLCs, and microemulsions) matrix for transdermal delivery was studied. Thereafter, this nanodispersion was transformed into the gel and evaluated for physicochemical attributes for dermal application. The *in vitro* skin permeation studies demonstrated better permeation through the skin *via* nanocarriers. Furthermore, skin irritation testing showed no toxicity. In addition to this, *in-vivo* studies exhibited significant improvement in cognitive functions [98].

Piperine, a natural alkaloid, was also investigated for therapeutic effects on AD. Obtained results were supportive as drug-loaded nanoparticles depicted finer effect in contrast to free drug and were able to restore the cognitive functions of the treated animals. Moreover, cytotoxicity studies highlighted the safety of the prepared formulations on the liver and kidneys. Furthermore, nanoformulations exhibited potential anti-inflammatory and anti-apoptotic activity of the loaded drugs, indicating the potential to stop AD progression [99]. A similar group tested piperine loaded chitosan nanoformulations in the treatment of AD [100]. In another similar study, piperine solid lipid nanoparticles were studied for targeting the brain *via* polysorbate-80 coating [101].

In a different study, Tabernaemontana divaricata an anticholinesterase alkaloidal extract loaded microstructures with a goal of skin permeation enhancement, was studied. For testing the hypothesis, the skin of a stillborn piglet was used. The authors further reported enhanced permeation resulted in elevated transdermal delivery of extract due to the microstructure carrier system [102].

Zhang *et al.* delivered Tet-1 Peptide conjugated selenium nanoparticles loaded epigallocatechin-3-gallate (EGCG) for anti-Alzheimer's therapy. *In-vitro* studies in PC12 cells revealed that Tet-1-EGCG@Se significantly inhibited Aβ aggregation and Aβ fibrillation with no signs of cytotoxicity and a high specific affinity for Aβ fibrils. Furthermore, it was observed that tagging the EGCG-modified SeNP with Tet-1 neuropeptide greatly enhanced *in vitro* neuronal targeting efficiency. Lastly, Tet-1 peptides can significantly enhance the cellular uptake of Tet-1-EGCG@Se in PC12 cells rather than in NIH/3T3 cells [103].

Recently, α-bisabolol loaded solid lipid nanoparticles were prepared by the hot homogenization method and evaluated for neuroprotective effect in Neuro2A cells. *in vitro*, antioxidant assays depicted that α-bisabolol-SLNs efficiently reduced oxidative stress by scavenging ROS production and improved neural growth along with anti-amyloidogenic agents. Furthermore, Th-T fluorescence assay elucidated that SLNs suppress the secondary structural alteration and formation of the toxic soluble oligomer, hence blocked Aβ aggregation by dissolving fibrils [104].

In another investigation, phytol fabricated PLGA nanoparticles were prepared by the solvent evaporation method and tested for cholinesterase inhibitory activity, anti-aggregation and disaggregation, free radical scavenging, and Th-T assay. The authors reported significant inhibition of AChE and BuChE activity, equivalent to standard drug donepezil. Additionally, DPPH radical scavenging assay for antioxidant activity revealed remarkable DPPH scavenging activity *vis-à-vis* pure drug suspension. Furthermore, phytol-PLGA NPs successfully disrupted amyloid aggregates and was non-cytotoxicity to Neuro2a cells [105].

Another study was conducted with a view of overcoming the gateway of the brain efficiently *via* the aid of a tailored delivery matrix, ultimately enhanced bioavailability. Herein, galantamine *via* intranasal administration contrary to oral delivery was studied by evaluating pharmacodynamic and biochemical parameters. The results of the pharmacodynamic studies and biochemical estimation of acetylcholinesterase activity in Swiss albino mice brain was found to be significant *via* intranasal delivery compared to oral administration. Moreover, significant recovery in amnesia was observed in the induced mice model by intranasal administration [106]. In a similar study, galantamine loaded chitosan nanoparticles were delivered intranasally in Wistar rats and were compared to oral delivery. The authors concluded that the polymeric nanoparticles successfully delivered the drug to the brain [107].

Furthermore, Gajbhiye *et al.* employed a new approach to deliver galantamine in a better way by conjugating Ascorbic acid to polymeric nanoparticles utilizing SVCT2 transporters of choroid plexus. The qualitative cellular uptake studies in NIT/3T3 cells inferred increased cellular uptake of drug targeted nanoparticles. Furthermore, *in-vivo* elucidations for pharmacodynamic demonstrated targeted nanoparticles significantly decreased escape latency and least working memory errors in contrast to free drug solution. Furthermore, AChE activity and higher accumulation of nanoformulations in the brain was observed contrary to pure drug suspension [108].

Huntington's Disease

Quercetin is a polyphenolic flavonoid, most widely found in fruits and vegetables. The antioxidant potential of quercetin has been studied from subcellular compartments to tissue levels in the brain. The neurodegeneration process initiates alongside the aging of the neurons. Despite common treatments that help to prevent the development of the disease, the condition of patients with progressive neurodegenerative diseases usually does not completely improve. Currently, the use of flavonoids, especially quercetin, for the treatment of neurodegenerative diseases, has been reviewed in animal models by Amanzadeh and coworkers [109]. However, the low bioavailability of quercetin has led researchers to construct various quercetin-involved nanoparticles. Undeniably, intranasal administration of quercetin nanoparticles, constructing superparamagnetic nanoparticles, and combinational treatment using nanoparticles such as quercetin and other drugs are suggested for future studies. In this subsection, we will discuss the delivery of the natural compound using nano-based strategies.

Intra- and extracellular protein aggregation is associated with a variety of neurodegenerative diseases. In a study, polylactide (PL)-based biodegradable nanoparticles have been explored to improve neuroprotection against polyglutamine (poly Q) aggregation, which is responsible for HD. PL is terminated with an anti-amyloidogenic trehalose molecule or the neurotransmitter dopamine, and then resultant nanoparticles are loaded with anti-amyloidogenic catechin molecules. The self-assembled nanoparticles entered into the neuronal cell and inhibited polyQ aggregation, lowered oxidative stress, and enhanced cell proliferation against polyQ aggregates [110].

In another work, the author aimed to load rosmarinic acid in solid lipid nanoparticles and evaluated it intranasally. The nasal delivery of nanoparticles produced significant therapeutic action as compared to intravenous application. *in vivo* biodistribution studies demonstrated higher brain drug concentration in comparison to its free counterpart. The encouraging results prove that lipidic carriers *via* non-invasive nose-to-brain drug delivery could be a promising therapeutic approach for effective management in HD [111].

Mitochondrial dysfunction is one of the key factors in HD pathogenesis. Therefore, treatment options to soothe mitochondrial impairments could provide a potential therapeutic intervention. Sandhir *et al.* formulated curcumin encapsulated solid lipid nanoparticles to improvise 3-nitropropionic acid (3-NP--induced HD in rats. Nonetheless, nanoformulations-treated animals showed a significant increase in the activity of mitochondrial complexes and cytochrome levels. Drug-loaded nanoparticles also restored the glutathione levels and superoxide dismutase activity. Additionally, a considerable decline in mitochondrial swelling, lipid peroxidation, protein carbonyls, and ROS was observed in rats treated with nano curcumin. Quantitative PCR and Western blot results displayed the activation of nuclear factor-erythroid 2 antioxidant pathways after nanoformulations administration in 3-NP-treated animals, along with improvement in neuromotor coordination [112].

Parkinson's Disease

Dopaminergic neurons of substantia nigra pars compacta are lost in Parkinson's due to α-Syn aggregation, and neurons are inflamed. In addition to this, the accumulation of α-Syn in the brain leads to the formation of Lewy bodies, and consequently, loss of dopaminergic neurons due to oxidative stress. To overcome these issues, Taebnia and coworkers formulated a drug carrier based on amine-functionalized mesoporous silica nanoparticles to study its effects on α-Syn fibrillation and cytotoxicity. Authors reported that nanoparticles interact strongly with the α-Syn species, ultimately causing inhibition of the fibrillation process [113].

Noninvasive delivery of focused ultrasound-assisted microbubbles induces the localized opening of BBB opening for better localization of therapeutics in the brain. On a similar concept, curcumin-loaded polysorbate 80-coated cerasomes were explored for the treatment of PD. The *in vitro* BBB cell-based assay showed cytocompatibility of cerasomes. Furthermore, *in vivo* permeability experiments revealed a significantly higher permeability of formulation through the BBB. The *in vivo* efficacy studies exhibited better therapeutic efficacy in a murine model [114]. In another study, combination therapy of curcumin with piperine loaded in lipidic carriers was explored [115]. Generally, rotenone induces neurotoxicity in dopaminergic cells is being widely studied in Parkinson's. In a study, authors synthesized curcumin loaded lactoferrin nanoparticles and evaluated them against rotenone-induced neurotoxicity. The authors further reported the improved biopharmaceutical properties of curcumin and pharmacodynamic profile in the rats [116]. In another study, the encapsulation of curcumin in alginate nanoparticles enhanced the neuroprotection through reducing oxidative stress and brain cell death in a transgenic Drosophila PD model [117].

Zhao and co-workers studied ferulic acid with an adipic acid linker and tannic acid as shell and core molecules to form nanoparticles *via* the flash nanoprecipitation method. The authors reported that nanoparticles showed a strong inhibitory effect on α-Syn fibrillization *in vitro* compared to untreated α-Syn using a Thioflavin T-assay [118].

Resveratrol, a natural antioxidant, shows a wide range of pharmacological actions, but its oral bioavailability is very low due to its extensive hepatic and presystemic metabolism. In an investigation, nanoemulsion showed high scavenging efficiency *in vitro*, and thereafter, *in vivo* pharmacokinetic studies revealed the higher concentration of the drug at the local site of action [119]. In a recent study, the protective effect of resveratrol nanoparticles against rotenone-induced neurodegeneration in rats was explored. The finding of the present study displayed the better efficacy of nanoparticles compared to free drug treatment in attenuating the rotenone-induced Parkinson's like behavioral alterations, biochemical and histological changes, oxidative stress, and mitochondrial dysfunction in rats [120]. In a similar study, it is concluded that resveratrol derived from *P. cuspidatum* and its liposomal form could protect the dopaminergic neurons in an experimental murine model of PD [121]. In addition to this, resveratrol-loaded nanocrystals exhibited favorable pharmacokinetics than its counterpart with higher plasma and brain concentrations. Finally, the authors concluded that the neuroprotective role of nanocrystals might be governed partially by Akt/Gsk3β signaling pathway [122]. Rocha and associates examine the neuroprotective effects of resveratrol-loaded polysorbate 80-coated poly(lactide) nanoparticles in a mouse model of PD. Furthermore, they showed remarkable neuroprotection against MPTP-induced behavioral and neurochemical changes [123].

Literature studies have made us believe that free radicals-induced neurodegeneration is the key factor in PD. Quercetin is a natural polyphenol that has been regarded as a significant player in changing the progression of neurodegenerative diseases by protecting damages caused by free radicals. Recently, the neuroprotective effects of quercetin nanocrystals on the 6-hydroxydopamine (6-OHDA)-induced Parkinson-like model in male rats was explored. Further, it established that nanocrystals prevented disturbance of memory, increased antioxidant enzyme activities, and total glutathione and reduced malondialdehyde levels in the hippocampal area [124]. In a similar investigation, authors formulated lecithin/cholesterol/2-hydroxypropyl-β-cyclodextrin nanosomes, Authors further proved that nanosomes of quercetin achieve higher brain concentrations and showed protective effects in experimental PD at the terminal field of lesioned substantia nigra neurons [125].

CONCLUSION

Neurodegenerative disorders are major health concerns nowadays. The treatment of NDs using conventional dosage forms faces a great barrier due to their poor biopharmaceutical attributes. Hence, the development of phytomolecules loaded nanoformulations can be a better strategy for a better therapeutic approach. Further, extensive studies on natural compound encapsulated nanoparticle delivery to the local site can improve the treatment of NDs. Their advantages, such as enhanced bioavailability, reduced dosing frequency, and targeted delivery, prove a strong ground for the management of NDs.

CONSENT FOR PUBLICATION

Not applicable.

CONFLICT OF INTEREST

The authors declare no conflict of interest, financial or otherwise.

ACKNOWLEDGEMENTS

The authors duly acknowledge ISF College of Pharmacy, Moga, Punjab for the motivation support.

REFERENCES

[1]   Sandhir R, Yadav A, Sunkaria A, Singhal N. Nano-antioxidants: An emerging strategy for intervention against neurodegenerative conditions. Neurochem Int 2015; 89: 209-26.
[http://dx.doi.org/10.1016/j.neuint.2015.08.011] [PMID: 26315960]

[2]   Ward RJ, Zucca FA, Duyn JH, Crichton RR, Zecca L. The role of iron in brain ageing and neurodegenerative disorders. Lancet Neurol 2014; 13(10): 1045-60.
[http://dx.doi.org/10.1016/S1474-4422(14)70117-6] [PMID: 25231526]

[3]   Shal B, Ding W, Ali H, Kim YS, Khan S. Anti-neuroinflammatory Potential of Natural Products in Attenuation of Alzheimer's Disease. Front Pharmacol 2018; 9: 548.
[http://dx.doi.org/10.3389/fphar.2018.00548] [PMID: 29896105]

[4]   Rasool M, Malik A, Qureshi MS, *et al.* Recent updates in the treatment of neurodegenerative disorders using natural compounds. Evid Based Complement Alternat Med 2014; 2014: 979730-7.
[http://dx.doi.org/10.1155/2014/979730] [PMID: 24864161]

[5]   Tysnes OB, Storstein A. Epidemiology of Parkinson's disease. J Neural Transm (Vienna) 2017; 124(8): 901-5.
[http://dx.doi.org/10.1007/s00702-017-1686-y] [PMID: 28150045]

[6]   Kumar A, Kumar V, Singh K, *et al.* Therapeutic Advances for Huntington's Disease. Brain Sci 2020; 10(1): 1-43.
[http://dx.doi.org/10.3390/brainsci10010043] [PMID: 31940909]

[7]   Jonson I, Ougland R, Larsen E. DNA Repair Mechanisms in Huntington's disease. Mol Neurobiol 47: 093-1102.
[http://dx.doi.org/10.1007/s12035-013-8409-7]

[8]   Xia R, Mao ZH. Progression of motor symptoms in Parkinson's disease. Neurosci Bull 2012; 28(1): 39-48.
[http://dx.doi.org/10.1007/s12264-012-1050-z] [PMID: 22233888]

[9]   Miller DB, O'Callaghan JP, Callaghan O. Biomarkers of Parkinson's disease: present and future. Metabolism 2015; 64(3) (Suppl. 1): S40-6.
[http://dx.doi.org/10.1016/j.metabol.2014.10.030] [PMID: 25510818]

[10]   Lee SG, Kim K, Kegelman TP, *et al.* Oncogene AEG-1 promotes glioma-induced neurodegeneration by increasing glutamate excitotoxicity. Cancer Res 2011; 71(20): 6514-23.
[http://dx.doi.org/10.1158/0008-5472.CAN-11-0782] [PMID: 21852380]

[11]   Sesti F, Liu S, Cai SQ. Oxidation of potassium channels by ROS: a general mechanism of aging and neurodegeneration? Trends Cell Biol 2010; 20(1): 45-51.
[http://dx.doi.org/10.1016/j.tcb.2009.09.008] [PMID: 19850480]

[12]   Patel RP, McAndrew J, Sellak H, *et al.* Biological aspects of reactive nitrogen species. Biochim Biophys Acta 1999; 1411(2-3): 385-400.
[http://dx.doi.org/10.1016/S0005-2728(99)00028-6] [PMID: 10320671]

[13]   Bano D, Dinsdale D, Cabrera-Socorro A, *et al.* Alteration of the nuclear pore complex in Ca^{2+})-mediated cell death. Cell Death Differ 2010; 17(1): 119-33.
[http://dx.doi.org/10.1038/cdd.2009.112] [PMID: 19713973]

[14]   Patricia GS, Sabine H. Autophagic pathology and calcium deregulation in neurodegeneration.in Current Topics in Neurotoxicity. J.M. Fuentes, Springer 2015; 9: 247-66.

[15]   Magi S, Castaldo P, Macrì ML, *et al.* Intracellular Calcium Dysregulation: Implications for Alzheimer's Disease. BioMed Res Int 2016; 2016: 6701324.
[http://dx.doi.org/10.1155/2016/6701324] [PMID: 27340665]

[16]   Velmurugan BK, Rathinasamy B, Lohanathan BP, Thiyagarajan V, Weng CF. Neuroprotective Role of Phytochemicals. Molecules 2018; 23(10): 1-15.
[http://dx.doi.org/10.3390/molecules23102485] [PMID: 30262792]

[17]   Aboul-Enein HY, Berczyńsk P, Kruk I. Phenolic compounds: the role of redox regulation in neurodegenerative disease and cancer. Mini Rev Med Chem 2013; 13(3): 385-98.
[PMID: 23190030]

[18]   Calcul L, Zhang B, Jinwal UK, Dickey CA, Baker BJ. Natural products as a rich source of tau-targeting drugs for Alzheimer's disease. Future Med Chem 2012; 4(13): 1751-61.
[http://dx.doi.org/10.4155/fmc.12.124] [PMID: 22924511]

[19]   Kumar GP, Khanum F. Neuroprotective potential of phytochemicals. Pharmacogn Rev 2012; 6(12): 81-90.
[http://dx.doi.org/10.4103/0973-7847.99898] [PMID: 23055633]

[20]   Kumar SV, Prabhu S, Rajalakhsmi S, *et al.* Review on potential phytocompounds in drug development for Parkinson disease: A pharmacoinformatic approach. Informatics in Medicine Unlocked 2016; 5: 15-25.
[http://dx.doi.org/10.1016/j.imu.2016.09.002]

[21]   Monroy A, Lithgow GJ, Alavez S. *Curcumin* and neurodegenerative diseases. Biofactors 2013; 39(1): 122-32.
[http://dx.doi.org/10.1002/biof.1063] [PMID: 23303664]

[22]   Luo Y. *Ginkgo biloba* neuroprotection: Therapeutic implications in Alzheimer's disease. J Alzheimers Dis 2001; 3(4): 401-7.
[http://dx.doi.org/10.3233/JAD-2001-3407] [PMID: 12214044]

[23]   Christen Y. Ginkgo biloba and neurodegenerative disorders. Front Biosci 2004; 9: 3091-104.
[http://dx.doi.org/10.2741/1462] [PMID: 15353340]

[24]   Huang X, Li N, Pu Y, Zhang T, Wang B. Neuroprotective Effects of *Ginseng* Phytochemicals: Recent Perspectives. Molecules 2019; 24(16): 2939.
[http://dx.doi.org/10.3390/molecules24162939] [PMID: 31416121]

[25]   Cho IH. Effects of *Panax ginseng* in Neurodegenerative Diseases. J Ginseng Res 2012; 36(4): 342-53.
[http://dx.doi.org/10.5142/jgr.2012.36.4.342] [PMID: 23717136]

[26]   Chaudhari KS, Tiwari NR, Tiwari RR, Sharma RS. Neurocognitive Effect of Nootropic Drug *Brahmi* (*Bacopa monnieri*) in Alzheimer's Disease. Ann Neurosci 2017; 24(2): 111-22.
[http://dx.doi.org/10.1159/000475900] [PMID: 28588366]

[27]   Kuboyama T, Tohda C, Komatsu K. Effects of Ashwagandha (roots of *Withania somnifera*) on neurodegenerative diseases. Biol Pharm Bull 2014; 37(6): 892-7.
[http://dx.doi.org/10.1248/bpb.b14-00022] [PMID: 24882401]

[28]   Amato A, Terzo S, Mulè F. Natural Compounds as Beneficial Antioxidant Agents in Neurodegenerative Disorders: A Focus on Alzheimer's Disease. Antioxidants 2019; 8(12): 608.
[http://dx.doi.org/10.3390/antiox8120608] [PMID: 31801234]

[29]   Ramassamy C. Emerging role of polyphenolic compounds in the treatment of neurodegenerative diseases: a review of their intracellular targets. Eur J Pharmacol 2006; 545(1): 51-64.
[http://dx.doi.org/10.1016/j.ejphar.2006.06.025] [PMID: 16904103]

[30]   Singh NA, Mandal AKA, Khan ZA. Potential neuroprotective properties of epigallocatechin-3-gallate (EGCG). Nutr J 2016; 15(1): 60.
[http://dx.doi.org/10.1186/s12937-016-0179-4] [PMID: 27268025]

[31]   Sun AY, Wang Q, Simonyi A, Sun GY. Resveratrol as a therapeutic agent for neurodegenerative diseases. Mol Neurobiol 2010; 41(2-3): 375-83.
[http://dx.doi.org/10.1007/s12035-010-8111-y] [PMID: 20306310]

[32]   Ayaz M, Sadiq A, Junaid M, *et al.* Flavonoids as Prospective Neuroprotectants and Their Therapeutic Propensity in Aging Associated Neurological Disorders. Front Aging Neurosci 2019; 11: 155.
[http://dx.doi.org/10.3389/fnagi.2019.00155] [PMID: 31293414]

[33]   de Andrade Teles RB, Diniz TC, Costa Pinto TC, *et al.* Flavonoids as Therapeutic Agents in Alzheimer's and Parkinson's Diseases: A Systematic Review of Preclinical Evidences. Oxid Med Cell Longev 2018; 2018: 7043213.
[http://dx.doi.org/10.1155/2018/7043213] [PMID: 29861833]

[34]   Ruszkowski P, Kozlowska TB. Natural Triterpenoids and their Derivatives with Pharmacological Activity Against Neurodegenerative Disorders. Mini Rev Org Chem 2014; 11: 307-15.
[http://dx.doi.org/10.2174/1570193X11031409151111559]

[35]   Ratheesh G, Tian L, Venugopal JR, *et al.* Role of medicinal plants in neurodegenerative diseases. Biomanuf Rev 2017; 2: 2-16.
[http://dx.doi.org/10.1007/s40898-017-0004-7]

[36]   Upadhyay RK. Drug delivery systems, CNS protection, and the blood brain barrier. BioMed Res Int 2014; 2014: 869269.
[http://dx.doi.org/10.1155/2014/869269] [PMID: 25136634]

[37]   Kang YJ, Cutler EG, Cho H. Therapeutic nanoplatforms and delivery strategies for neurological disorders. Nano Converg 2018; 5(1): 35.
[http://dx.doi.org/10.1186/s40580-018-0168-8] [PMID: 30499047]

[38]   Jeevanandam J, Barhoum A, Chan YS, Dufresne A, Danquah MK. Review on nanoparticles and nanostructured materials: history, sources, toxicity and regulations. Beilstein J Nanotechnol 2018; 9: 1050-74.
[http://dx.doi.org/10.3762/bjnano.9.98] [PMID: 29719757]

[39]   Niu X, Chen J, Gao J. Nanocarriers as a powerful vehicle to overcome blood-brain barrier in treating

neurodegenerative diseases: Focus on recent advances. Asian J Pharm Sci 2019; 14(5): 480-96.
[http://dx.doi.org/10.1016/j.ajps.2018.09.005] [PMID: 32104476]

[40]   Patra JK, Das G, Fraceto LF, *et al.* Nano based drug delivery systems: recent developments and future prospects. J Nanobiotechnology 2018; 16(1): 71.
[http://dx.doi.org/10.1186/s12951-018-0392-8] [PMID: 30231877]

[41]   Wong HL, Wu XY, Bendayan R. Nanotechnological advances for the delivery of CNS therapeutics. Adv Drug Deliv Rev 2012; 64(7): 686-700.
[http://dx.doi.org/10.1016/j.addr.2011.10.007] [PMID: 22100125]

[42]   Md S, Mustafa G, Baboota S, Ali J. Nanoneurotherapeutics approach intended for direct nose to brain delivery. Drug Dev Ind Pharm 2015; 41(12): 1922-34.
[http://dx.doi.org/10.3109/03639045.2015.1052081] [PMID: 26057769]

[43]   Wilczewska AZ, Niemirowicz K, Markiewicz KH, Car H. Nanoparticles as drug delivery systems. Pharmacol Rep 2012; 64(5): 1020-37.
[http://dx.doi.org/10.1016/S1734-1140(12)70901-5] [PMID: 23238461]

[44]   Durgavati Yadav D, Sandeep K, Pandey D, *et al.* Liposomes for Drug Delivery. J Biotechnol Biomater 2017; 7(4): 2-8.

[45]   Cunha S, Amaral MH, Lobo JMS, Silva AC. Lipid nanoparticles for nasal/intranasal drug delivery. Crit Rev Ther Drug Carrier Syst 2017; 34(3): 257-82.
[http://dx.doi.org/10.1615/CritRevTherDrugCarrierSyst.2017018693] [PMID: 28845761]

[46]   Li Q, Cai T, Huang Y, Xia X, Cole SPC, Cai Y. A Review of the structure, preparation, and application of NLCs, PNPs, and PLNs. Nanomaterials (Basel) 2017; 7(6): 122.
[http://dx.doi.org/10.3390/nano7060122] [PMID: 28554993]

[47]   Reddy LH, Sharma RK, Chuttani K, Mishra AK, Murthy RR. Etoposide-incorporated tripalmitin nanoparticles with different surface charge: formulation, characterization, radiolabeling, and biodistribution studies. AAPS J 2004; 6(3): e23.
[http://dx.doi.org/10.1208/aapsj060323] [PMID: 15760108]

[48]   Ko YT, Bhattacharya R, Bickel U. Liposome encapsulated polyethylenimine/ODN polyplexes for brain targeting. J Control Release 2009; 133(3): 230-7.
[http://dx.doi.org/10.1016/j.jconrel.2008.10.013] [PMID: 19013203]

[49]   Nasir S, Hussein MZ, Zainal Z, Yusof NA. Carbon-Based Nanomaterials/Allotropes: A Glimpse of Their Synthesis, Properties and Some Applications. Materials (Basel) 2018; 11(2): 2-24.
[http://dx.doi.org/10.3390/ma11020295] [PMID: 29438327]

[50]   Kumar P, Modi G, Choonara YE. Nano(neuro) medicinal interventions for neurodegenerative disorders: a meta-analysis of concurrent challenges and strategic solutions. Nanotechnology and Drug Delivery JL Arias. 2016; 1: 284.

[51]   Lohan S, Raza K, Mehta SK, Bhatti GK, Saini S, Singh B. Anti-Alzheimer's potential of berberine using surface decorated multi-walled carbon nanotubes: A preclinical evidence. Int J Pharm 2017; 530(1-2): 263-78.
[http://dx.doi.org/10.1016/j.ijpharm.2017.07.080] [PMID: 28774853]

[52]   John AA, Subramanian AP, Vellayappan MV, Balaji A, Mohandas H, Jaganathan SK. Carbon nanotubes and graphene as emerging candidates in neuroregeneration and neurodrug delivery. Int J Nanomedicine 2015; 10: 4267-77.
[PMID: 26170663]

[53]   Dugan LL, Lovett EG, Quick KL, Lotharius J, Lin TT, O'Malley KL. Fullerene-based antioxidants and neurodegenerative disorders. Parkinsonism Relat Disord 2001; 7(3): 243-6.
[http://dx.doi.org/10.1016/S1353-8020(00)00064-X] [PMID: 11331193]

[54]   Alawdi SH, El-Denshary ES, Safar MM, Eidi H, David MO, Abdel-Wahhab MA. Neuroprotective effect of nanodiamond in Alzheimer's disease rat model: a pivotal role for modulating NF-κB and

STAT3 signaling. Mol Neurobiol 2017; 54(3): 1906-18.
[http://dx.doi.org/10.1007/s12035-016-9762-0] [PMID: 26897372]

[55]   Haziza S, Mohan N, Loe-Mie Y, *et al.* Fluorescent nanodiamond tracking reveals intraneuronal transport abnormalities induced by brain-disease-related genetic risk factors. Nat Nanotechnol 2017; 12(4): 322-8.
[http://dx.doi.org/10.1038/nnano.2016.260] [PMID: 27893730]

[56]   Tosi G, Costantino L, Ruozi B, Forni F, Vandelli MA. Polymeric nanoparticles for the drug delivery to the central nervous system. Expert Opin Drug Deliv 2008; 5(2): 155-74.
[http://dx.doi.org/10.1517/17425247.5.2.155] [PMID: 18248316]

[57]   Liechty WB, Kryscio DR, Slaughter BV, Peppas NA. Polymers for drug delivery systems. Annu Rev Chem Biomol Eng 2010; 1: 149-73.
[http://dx.doi.org/10.1146/annurev-chembioeng-073009-100847] [PMID: 22432577]

[58]   Costantino L, Boraschi D. Is there a clinical future for polymeric nanoparticles as brain-targeting drug delivery agents? Drug Discov Today 2012; 17(7-8): 367-78.
[http://dx.doi.org/10.1016/j.drudis.2011.10.028] [PMID: 22094246]

[59]   Georgieva JV, Hoekstra D, Zuhorn IS. Smuggling drugs into the brain: an overview of ligands targeting transcytosis for drug delivery across the blood–brain barrier. Pharmaceutics 2014; 6(4): 557-83.
[http://dx.doi.org/10.3390/pharmaceutics6040557] [PMID: 25407801]

[60]   Benseny-Cases N, Klementieva O, Cladera J. Dendrimers antiamyloidogenic potential in neurodegenerative diseases. New J Chem 2012; 36: 211-6.
[http://dx.doi.org/10.1039/C1NJ20469F]

[61]   Mignani S, Bryszewska M, Zablocka M, *et al.* Can dendrimer based nanoparticles fight neurodegenerative diseases? Current situation *versus* other established approaches. Prog Polym Sci 2017; 64: 23-51.
[http://dx.doi.org/10.1016/j.progpolymsci.2016.09.006]

[62]   Gangu KK, Maddila S, Saratchandra Babu Mukkamala SB, *et al.* A review on contemporary Metal–Organic Framework materials. Inorg Chim Acta 2016; 446: 61-74.
[http://dx.doi.org/10.1016/j.ica.2016.02.062]

[63]   Zhu QL, Xu Q. Metal-organic framework composites. Chem Soc Rev 2014; 43(16): 5468-512.
[http://dx.doi.org/10.1039/C3CS60472A] [PMID: 24638055]

[64]   Beg S, Rahman M, Jain A, *et al.* Nanoporous metal organic frameworks as hybrid polymer-metal composites for drug delivery and biomedical applications. Drug Discov Today 2017; 22(4): 625-37.
[http://dx.doi.org/10.1016/j.drudis.2016.10.001] [PMID: 27742533]

[65]   Qu W, Zuo W, Li N, *et al.* Design of multifunctional liposome-quantum dot hybrid nanocarriers and their biomedical application. J Drug Target 2017; 25(8): 661-72.
[http://dx.doi.org/10.1080/1061186X.2017.1323334] [PMID: 28438041]

[66]   Huang N, Cheng S, Zhang X, *et al.* Efficacy of NGR peptide-modified PEGylated quantum dots for crossing the blood-brain barrier and targeted fluorescence imaging of glioma and tumor vasculature. Nanomedicine (Lond) 2017; 13(1): 83-93.
[http://dx.doi.org/10.1016/j.nano.2016.08.029] [PMID: 27682740]

[67]   Pattni BS, Chupin VV, Torchilin VP. New developments in liposomal drug delivery. Chem Rev 2015; 115(19): 10938-66.
[http://dx.doi.org/10.1021/acs.chemrev.5b00046] [PMID: 26010257]

[68]   Kassem AA, Abd El-Alim SH, Basha M, Salama A. Phospholipid complex enriched micelles: A novel drug delivery approach for promoting the antidiabetic effect of repaglinide. Eur J Pharm Sci 2017; 99: 75-84.
[http://dx.doi.org/10.1016/j.ejps.2016.12.005] [PMID: 27998799]

[69] Vauthier C, Bouchemal K. Methods for the preparation and manufacture of polymeric nanoparticles. Pharm Res 2009; 26(5): 1025-58.
[http://dx.doi.org/10.1007/s11095-008-9800-3] [PMID: 19107579]

[70] Reis CP, Neufeld RJ, Ribeiro AJ, Veiga F. Nanoencapsulation I. Methods for preparation of drug-loaded polymeric nanoparticles. Nanomedicine (Lond) 2006; 2(1): 8-21.
[http://dx.doi.org/10.1016/j.nano.2005.12.003] [PMID: 17292111]

[71] Wang Y, Li P, Truong-Dinh Tran T, Zhang J, Kong L. Manufacturing Techniques and Surface Engineering of Polymer Based Nanoparticles for Targeted Drug Delivery to Cancer. Nanomaterials (Basel) 2016; 6(2): 26.
[http://dx.doi.org/10.3390/nano6020026] [PMID: 28344283]

[72] Pal SL, Jana Manna UP, et al. Nanoparticle: An overview of preparation and characterization. J Appl Pharm Sci 2011; 06: 228-34.

[73] Byrappa K, Ohara S, Adschiri T. Nanoparticles synthesis using supercritical fluid technology-towards biomedical applications. Adv Drug Deliv Rev 2008; 60(3): 299-327.
[http://dx.doi.org/10.1016/j.addr.2007.09.001] [PMID: 18192071]

[74] Sheth P, Sandhu H, Singhal D, Malick W, Shah N, Kislalioglu MS. Nanoparticles in the pharmaceutical industry and the use of supercritical fluid technologies for nanoparticle production. Curr Drug Deliv 2012; 9(3): 269-84.
[http://dx.doi.org/10.2174/156720112800389052] [PMID: 22283656]

[75] Tyagi S, Pandey VK. Research and Reviews: Journal of Pharmaceutics and Nanotechnology. JPN 2016; 4: 2-12.

[76] Liu Y, Zhou H, Yin T, et al. Quercetin-modified gold-palladium nanoparticles as a potential autophagy inducer for the treatment of Alzheimer's disease. J Colloid Interface Sci 2019; 552: 388-400.
[http://dx.doi.org/10.1016/j.jcis.2019.05.066] [PMID: 31151017]

[77] Pinhero RGR, Granja A. Quercetin lipid nanoparticles finctionalized with transferin for Alzheimer's disease. Jejps 2020; 105314.

[78] Sun D, Li N, Zhang W, et al. Design of PLGA-functionalized quercetin nanoparticles for potential use in Alzheimer's disease. Colloids Surf B Biointerfaces 2016; 148: 116-29.
[http://dx.doi.org/10.1016/j.colsurfb.2016.08.052] [PMID: 27591943]

[79] Palle S, Neerati P. Quercetin nanoparticles attenuates scopolamine induced spatial memory deficits and pathological damages in rats. Jbfopcu 2016; 10(004): 101-6.

[80] Almendra E G. Effect of the oral administration of nano encapsulated quercetin on a mouse model of Alzheimer's disease. Jijpharm 2016; 11(61): 50-7.

[81] Nday CM, Eleftheriadou D, Jackson G. Naringin nanoparticles against neurodegenerative processes: A preliminary work. Hell J Nucl Med 2019; 22 (Suppl.): 32-41.
[PMID: 30877721]

[82] Kheradmand E, Hajizadeh Moghaddam A, Zare M. Neuroprotective effect of hesperetin and nano-hesperetin on recognition memory impairment and the elevated oxygen stress in rat model of Alzheimer's disease. Biomed Pharmacother 2018; 97: 1096-101.
[http://dx.doi.org/10.1016/j.biopha.2017.11.047] [PMID: 29136946]

[83] Lu X, Ji C, Xu H, et al. Resveratrol-loaded polymeric micelles protect cells from Abeta-induced oxidative stress. Int J Pharm 2009; 375(1-2): 89-96.
[http://dx.doi.org/10.1016/j.ijpharm.2009.03.021] [PMID: 19481694]

[84] Loureiro JA, Andrade S, Duarte A, et al. Resveratrol and Grape Extract-loaded Solid Lipid Nanoparticles for the Treatment of Alzheimer's Disease. Molecules 2017; 22(2): 1-16.
[http://dx.doi.org/10.3390/molecules22020277] [PMID: 28208831]

[85]   Frozza L R, Salbego C. Incorporation of resveratrol into lipid-core nanocapsules improves its cerebral bioavailability and reduces the ab-induced toxicity. Jjalz 2011; 7: S114-4.

[86]   Yang L, Wang W, Chen J, Wang N, Zheng G. A comparative study of resveratrol and resveratrol-functional selenium nanoparticles: Inhibiting amyloid β aggregation and reactive oxygen species formation properties. J Biomed Mater Res A 2018; 106(12): 3034-41.
[http://dx.doi.org/10.1002/jbm.a.36493] [PMID: 30295993]

[87]   Bondì ML, Montana G. Ferulic acid-loaded lipid nanostructures as drug delivery systems for alzheimer's disease: preparation, characterization and cytotoxicity studies. Curr Nanosci 2009; 5: 26-32.
[http://dx.doi.org/10.2174/157341309787314656]

[88]   Cheng KK, Yeung CF, Ho SW, Chow SF, Chow AH, Baum L. Highly stabilized curcumin nanoparticles tested in an *in vitro* blood-brain barrier model and in Alzheimer's disease Tg2576 mice. AAPS J 2013; 15(2): 324-36.
[http://dx.doi.org/10.1208/s12248-012-9444-4] [PMID: 23229335]

[89]   Mathew A, Fukuda T. Curcumin Loaded-PLGA Nanoparticles Conjugated withTet-1 Peptide for Potential Use in Alzheimer's disease. Plos One 2013; 7(3): e32616. 1-10

[90]   Shankar N, Spenner C. Potential Treatment for Alzheimer's Disease:Encapsulation of Curcumin in Polymeric PLGAPEGNanoparticles Rescues Neuro2A cellsfrom Beta-Amyloid Induced Cytotoxicity andCaspase Induced Apoptosis. JE SS 2016; 3: 4.

[91]   Huang N, Lu S, Liu XG, Zhu J, Wang YJ, Liu RT. PLGA nanoparticles modified with a BBB-penetrating peptide co-delivering Aβ generation inhibitor and curcumin attenuate memory deficits and neuropathology in Alzheimer's disease mice. Oncotarget 2017; 8(46): 81001-13.
[http://dx.doi.org/10.18632/oncotarget.20944] [PMID: 29113362]

[92]   Fan S, Zheng Y, Liu X, *et al.* Curcumin-loaded PLGA-PEG nanoparticles conjugated with B6 peptide for potential use in Alzheimer's disease. Drug Deliv 2018; 25(1): 1091-102.
[http://dx.doi.org/10.1080/10717544.2018.1461955] [PMID: 30107760]

[93]   Huo X, Zhang Y. A novel synthesis of selenium nanoparticles encapsulated PLGAnanospheres with curcumin molecules for the inhibition of amyloid β aggregation in Alzheimer's disease. JPhotoBiol 2018; 98-102. S1011-1344,(18),31165-5

[94]   Kuo YC, Tsai HC. Rosmarinic acid- and curcumin-loaded polyacrylamide-cardiolipin-poly(lactide-co-glycolide) nanoparticles with conjugated 83-14 monoclonal antibody to protect β-amyloid-insulted neurons. Mater Sci Eng C 2018; 91: 445-457, 445-457.
[http://dx.doi.org/10.1016/j.msec.2018.05.062] [PMID: 30033276]

[95]   Mathew A, Fukuda T. Curcumin-conjugated nanoliposomes with high affinity for Aβ deposits: Possible applications to Alzheimer disease. Journalpone 2013; 7: 1-10.

[96]   Mourtas S, Canovi M, Zona C, *et al.* Curcumin-decorated nanoliposomes with very high affinity for amyloid-β1-42 peptide. Biomaterials 2011; 32(6): 1635-45.
[http://dx.doi.org/10.1016/j.biomaterials.2010.10.027] [PMID: 21131044]

[97]   Meng Q, Wang A, Hua H, *et al.* Intranasal delivery of Huperzine A to the brain using lactoferrin-conjugated N-trimethylated chitosan surface-modified PLGA nanoparticles for treatment of Alzheimer's disease. Int J Nanomedicine 2018; 13: 705-18.
[http://dx.doi.org/10.2147/IJN.S151474] [PMID: 29440896]

[98]   Patel PA, Patil SC, Kalaria DR, Kalia YN, Patravale VB. Comparative *in vitro* and *in vivo* evaluation of lipid based nanocarriers of Huperzine A. Int J Pharm 2013; 446(1-2): 16-23.
[http://dx.doi.org/10.1016/j.ijpharm.2013.02.014] [PMID: 23410989]

[99]   Elnaggar S RY, Etman SM. Novel piperine-loaded Tween-integrated monooleincubosomes as brain-targeted oral nanomedicine in Alzheimer's disease: pharmaceutical, biological, and toxicological studies. IJN 2015; 10: 5459-73.

[100] Elnaggar SRY. ETMAN M.S.Intranasal Piperine-Loaded Chitosan Nanoparticles as Brain-Targeted Therapy in Alzheimer's disease: Optimization, Biological Efficacy, and Potential Toxicity. J Pharm Sci 2015; 104: 3544-56.
[http://dx.doi.org/10.1002/jps.24557]

[101] Yusuf M, Khan M, Khan RA, Ahmed B. Preparation, characterization, *in vivo* and biochemical evaluation of brain targeted Piperine solid lipid nanoparticles in an experimentally induced Alzheimer's disease model. J Drug Target 2013; 21(3): 300-11.
[http://dx.doi.org/10.3109/1061186X.2012.747529] [PMID: 23231324]

[102] Chaiyana W, Rades T, Okonogi S. Characterization and *in vitro* permeation study of microemulsions and liquid crystalline systems containing the anticholinesterase alkaloidal extract from Tabernaemontana divaricata. Int J Pharm 2013; 452(1-2): 201-10.
[http://dx.doi.org/10.1016/j.ijpharm.2013.05.005] [PMID: 23680734]

[103] Zhang J, Zhou X, Yu Q, *et al.* Epigallocatechin-3-gallate (EGCG)-stabilized selenium nanoparticles coated with Tet-1 peptide to reduce amyloid-β aggregation and cytotoxicity. ACS Appl Mater Interfaces 2014; 6(11): 8475-87.
[http://dx.doi.org/10.1021/am501341u] [PMID: 24758520]

[104] Sathya S, Shanmuganathan B. α-Bisabolol loaded solid lipid nanoparticles attenuates Aβ aggregation and protects Neuro2A cells from Aβ induced neurotoxicity. J molliq 2018; 05: 075.

[105] Sathya S, Shanmuganathan B, Saranya S, Vaidevi S, Ruckmani K, Pandima Devi K. Phytol-loaded PLGA nanoparticle as a modulator of Alzheimer's toxic Aβ peptide aggregation and fibrillation associated with impaired neuronal cell function. Artif Cells Nanomed Biotechnol 2018; 46(8): 1719-30.
[PMID: 29069924]

[106] Sunena , Singh SK, Mishra DN. Sunena; Singh. K. Nose to brain delivery of galantamine loaded nanoparticles: *in-vivo* pharmacodynamic and biochemical study in mice. Curr Drug Deliv 2019; 16(1): 51-8.
[http://dx.doi.org/10.2174/1567201815666181004094707] [PMID: 30289074]

[107] Hanafy AS, Farid RM, Helmy MW, ElGamal SS. Pharmacological, toxicological and neuronal localization assessment of galantamine/chitosan complex nanoparticles in rats: future potential contribution in Alzheimer's disease management. Drug Deliv 2016; 23(8): 3111-22.
[http://dx.doi.org/10.3109/10717544.2016.1153748] [PMID: 26942549]

[108] Gajbhiye KR, Gajbhiye V, Siddiqui IA, Pilla S, Soni V. Ascorbic acid tethered polymeric nanoparticles enable efficient brain delivery of galantamine: An *in vitro-in vivo* study. Sci Rep 2017; 7(1): 11086.
[http://dx.doi.org/10.1038/s41598-017-11611-4] [PMID: 28894228]

[109] Amanzadeh E, Esmaeili A, Rahgozar S, Nourbakhshnia M. Application of quercetin in neurological disorders: from nutrition to nanomedicine. Rev Neurosci 2019; 30(5): 555-72.
[http://dx.doi.org/10.1515/revneuro-2018-0080] [PMID: 30753166]

[110] Mandal S, Debnath K, Jana NR, Jana NR. Trehalose Conjugated, Catechin Loaded Polylactide Nanoparticle for Improved Neuroprotection against Intracellular Polyglutamine Aggregate. Biomacromolecules 2020; 1-27.
[http://dx.doi.org/10.1021/acs.biomac.0c00143]

[111] Bhatt R, Singh D, Prakash A, Mishra N. Development, characterization and nasal delivery of rosmarinic acid-loaded solid lipid nanoparticles for the effective management of Huntington's disease. Drug Deliv 2015; 22(7): 931-9.
[http://dx.doi.org/10.3109/10717544.2014.880860] [PMID: 24512295]

[112] Sandhir R, Yadav A, Mehrotra A, Sunkaria A, Singh A, Sharma S. Curcumin nanoparticles attenuate neurochemical and neurobehavioral deficits in experimental model of Huntington's disease. Neuromolecular Med 2014; 16(1): 106-18.

[http://dx.doi.org/10.1007/s12017-013-8261-y] [PMID: 24008671]

[113]   Taebnia N, Morshedi D, Yaghmaei S, Aliakbari F, Rahimi F, Arpanaei A. Curcumin-loaded amine-functionalized mesoporous silica nanoparticles inhibit α-synuclein fibrillation and reduce its cytotoxicity-associated effects. Langmuir 2016; 32(50): 13394-402.
[http://dx.doi.org/10.1021/acs.langmuir.6b02935] [PMID: 27993021]

[114]   Zhang N, Yan F, Liang X, *et al.* Localized delivery of curcumin into brain with polysorbate 80-modified cerasomes by ultrasound-targeted microbubble destruction for improved Parkinson's disease therapy. Theranostics 2018; 8(8): 2264-77.
[http://dx.doi.org/10.7150/thno.23734] [PMID: 29721078]

[115]   Kundu P, Das M, Tripathy K, Sahoo SK. Delivery of dual drug loaded lipid based nanoparticles across the blood–brain barrier impart enhanced neuroprotection in a rotenone induced mouse model of Parkinson's disease. ACS Chem Neurosci 2016; 7(12): 1658-70.
[http://dx.doi.org/10.1021/acschemneuro.6b00207] [PMID: 27642670]

[116]   Bollimpelli VS, Kumar P, Kumari S, Kondapi AK. Neuroprotective effect of curcumin-loaded lactoferrin nano particles against rotenone induced neurotoxicity. Neurochem Int 2016; 95: 37-45.
[http://dx.doi.org/10.1016/j.neuint.2016.01.006] [PMID: 26826319]

[117]   Siddique YH, Khan W, Singh BR, Naqvi AH. Synthesis of alginate-curcumin nanocomposite and its protective role in transgenic Drosophila model of Parkinson's disease. ISRN Pharmacol 2013; 2013: 794582.
[http://dx.doi.org/10.1155/2013/794582] [PMID: 24171120]

[118]   Zhao N, Yang X, Calvelli HR, *et al.* Antioxidant nanoparticles for concerted inhibition of α-synuclein fibrillization, and attenuation of microglial intracellular aggregation and activation. Front Bioeng Biotechnol 2020; 8: 112.
[http://dx.doi.org/10.3389/fbioe.2020.00112] [PMID: 32154238]

[119]   Pangeni R, Sharma S, Mustafa G, Ali J, Baboota S. Vitamin E loaded resveratrol nanoemulsion for brain targeting for the treatment of Parkinson's disease by reducing oxidative stress. Nanotechnology 2014; 25(48): 485102-14.
[http://dx.doi.org/10.1088/0957-4484/25/48/485102] [PMID: 25392203]

[120]   Palle S, Neerati P. Improved neuroprotective effect of resveratrol nanoparticles as evinced by abrogation of rotenone-induced behavioral deficits and oxidative and mitochondrial dysfunctions in rat model of Parkinson's disease. Naunyn Schmiedebergs Arch Pharmacol 2018; 391(4): 445-53.
[http://dx.doi.org/10.1007/s00210-018-1474-8] [PMID: 29411055]

[121]   Wang Y, Xu H, Fu Q, Ma R, Xiang J. Protective effect of resveratrol derived from Polygonum cuspidatum and its liposomal form on nigral cells in parkinsonian rats. J Neurol Sci 2011; 304(1-2): 29-34.
[http://dx.doi.org/10.1016/j.jns.2011.02.025] [PMID: 21376343]

[122]   Xiong S, Liu W, Zhou Y, *et al.* Enhancement of oral bioavailability and anti-Parkinsonian efficacy of resveratrol through a nanocrystal formulation. Asian Journal of Pharmaceutical Sciences 2019; 1-11.
[http://dx.doi.org/10.1016/j.ajps.2019.04.003]

[123]   da Rocha Lindner G, Bonfanti Santos D, Colle D, *et al.* da. Improved neuroprotective effects of resveratrol-loaded polysorbate 80-coated poly(lactide) nanoparticles in MPTP-induced Parkinsonism. Nanomedicine (Lond) 2015; 10(7): 1127-38.
[http://dx.doi.org/10.2217/nnm.14.165] [PMID: 25929569]

[124]   Ghaffari F, Hajizadeh Moghaddam A, Zare M. Neuroprotective effect of quercetin nanocrystal in a 6-hydroxydopamine model of parkinson disease: biochemical and behavioral evidence. Basic Clin Neurosci 2018; 9(5): 317-24.
[http://dx.doi.org/10.32598/bcn.9.5.317] [PMID: 30719246]

[125]   Díaz M, Vaamonde L, Dajas F. Assessment of the protective capacity of nanosomes of quercetin in an experimental model of parkinsons disease in the rat. General Medicine: Open Access 2015; 1-7.

CHAPTER 10

Recent Advancement in the Nanoparticles Mediated Therapeutics of Parkinson's Disease

Vivek K. Chaturvedi[1], Payal Singh[2] and M.P. Singh[1,*]

[1] *Centre of Biotechnology, University of Allahabad, Prayagraj-221002, India*

[2] *Department of Zoology, MMV, Banaras Hindu University, Varanasi-221005, India*

Abstract: Nanoparticle plays a very effective role in the therapeutics of Parkinson's disease (PD). The blood-brain barrier (BBB) is the main barrier that prevents the efficiency of any therapeutic compound.Nanoparticles overcome this problem by crossing the BBB. Recently many nanoparticles show promising responses in PD. Silver, gold, and many other nanoparticles effectively prevent progressive neurodegeneration in PD. In this book chapter, we have included some recent development in the nanoparticles mediated therapeutics for PD.

Keywords: Gold Nanoparticle, Growth Factor, Nanoparticles, Parkinson's disease.

INTRODUCTION

Parkinson's disease (PD) is the second most common neurodegenerative disease of the aging population after Alzheimer's disease, which is characterized by specific motor as well as some nonmotor symptoms [1, 2]. Resting tremors, bradykinesia, postural imbalance, *etc.*, are the most common motor symptoms, while constipation, sleeping abnormalities, fatigue, *etc.* considered the nonmotor ones [3, 4]. Till now, the complete cure for this progressive disease is not possible; levodopa (L-Dopa) dependent therapy is one of the best treatments available that only use to manage the symptomatic response that ultimately cause L-Dopa induced dyskinesia upon prolonged use [5, 6]. In the last few decades, researchers have tried several alternative ways to combat this disease effectively. Herbal mediated therapy is the best example of PD therapy having minimum side effects [7]. *Mucuna pruriens, Withania somnifera, Tinospora cordifolia, etc.* are some common examples of herbal plant that shows the promising response in

---

* **Corresponding author M.P. Singh:** Centre of Biotechnology, University of Allahabad, Prayagraj-221002, India; Tel: +91 9415677998; E-mail: mpsingh.16@gmail.com

Sachchida Nand Rai (Ed.)

PD treatment with minimum side effect [8 - 12]. Some chemical compounds like Ursolic acid and Chlorogenic acid also exert significant neuroprotection for PD [13 - 15]. The blood-brain barrier (BBB) is the biggest culprit that minimizes the therapeutic response of the above-mentioned herbal plant and chemicals. Researchers have made an enormous attempt to resolve this barrier problem. Nanoparticles are the best way to overcome this problem because of the smallest size that easily penetrates the BBB and also causes minimal side effects with greater efficiency [16, 17]. Following this, we have discussed the nanoparticles-based therapy PD in the following paragraph.

NANOPARTICLES BASED THERAPY FOR PARKINSON'S DISEASE

Gold nanoclusters exhibited a strong therapeutic response in the Parkinsonian mouse model. In this study, the authors wanted to optimize the route for these gold nanoclusters (AuNCs) in the PD model. Different routes of administration, for example, intravenous, intraperitoneal, gavage, and intranasal injection, wereutilizedto find the efficacy of these nanoclusters. Different microscopies were usedto check the biodistribution of AuNCs in mice. Both *in vitro*(cells) and *in vivo* (mice) were used to assess the toxicity of AuNCs. In the mouse brain, intraperitoneal administration showed better bioavailability and half-life as compared to other routes. AuNCs easily crossed the BBB and were excreted by the kidney, and no toxicity was found in both cells and mice regarding these nanoclusters. Authors have concluded that the intraperitoneal route is the most effective one, which shows a better response in comparison with other routes of entry [18]. In the mouse model of PD, dextran-coated iron oxide nanoparticle improves the efficacy of human mesenchymal stem cells in the manifold. This stem cell can differentiate into dopaminergic neurons and also secrets various neurotrophic factors. Mesenchymal stem cells exhibit the ability to replace damaged neurons. Therefore, we can say that this dextran-coated iron oxide nanoparticle prevents the progressive degeneration of dopaminergic neurons indirectly by improving the function of mesenchymal stem cells [19]. Cerium oxide nanoparticles(CeO2NPs) showed the anti-PD potential in the 6-hydroxydopamine (6-OHDA) induced Parkinsonian rat model. Behavioral impairment was induced by this PD toxin OHDA which wassignificantly improved by these nanoparticles as demonstrated by an open field, rotarod, and stepping test. Also, CeO2NPs reversed the neurobiochemical abnormalities in the striatum. CeO2NPs decreased apoptosis and slightly ameliorated the dopamine level in the stratum. The authors have concluded the antiapoptotic and antioxidative activity in the 6-OHDA intoxicated model of PD [20]. Further study will be needed to confirm the same.

Growth factor (GF) based therapy shows a very promising response for the PD treatment [21]. Among different GF, glial cell-derived neurotrophic factor (GDNF) is the most studied one and also passed the double-blind clinical trial for PD [22]. GDNF's shorter half-life, rapid degradation rate, and limited capacity to cross the BBB reduces its success in clinical studies. One interesting study overcomes this BBB and other hurdles of GDNF. In this study, GDNF was encapsulated into chitosan (CS)-coated nanostructured lipid carriers, with the surface-modified with transactivator of transcription (TAT) peptide (CS-nanostructured lipid carrier (NLC)-TAT-GDNF) and administered *via* the intranasal route in 1-methyl-4-phenyl-1,2,3,6-tetrahydropyridine (MPTP) intoxicated mouse model of PD. This nanostructure improved the motor symptoms in the Parkinsonian model. In addition, the count of marker enzyme of PD pathogenesis, which is tyrosine hydroxylase (TH), also improved in striatum and substantia nigra by this GDNF coated nanostructure. This CS-NLC-T-T-GDNF can reverse the Parkinsonian symptoms and suggest an alternative and effective therapeutic option for PD [23].

Gold Nanoparticles (GNP) show a considerable amount of neuroprotective activity in the Parkinsonian model and *in vitro*. A study showed the potential of GNP on PC12 cells and in mice. GNP inhibited the process of apoptosis in PC12 cells and of dopaminergic neurons. A significant amount of GNP in the brain of the mouse showed that GNP effectively crosses the BBB. GNP showed potent neuroprotective activity in both model systems [24]. In 6-OHDA partially lesioned rats' model of PD, vascular endothelial growth factor (VEGF) and GDNF-loaded nanospheres show strong synergistic Anti-Parkinsonian activity. This synergistic combination might be used as an alternative option for PD therapy [25]. For the efficient intranasal delivery of rotigotine in the PD model, a study used biodegradable poly(ethylene glycol)–poly(lactic-co-glycolic acid) (PEG-PLGA) nanoparticles (NPs). These NPs effectively targeted the rotigotine into the brain, as suggested by the intranasal administration. This study explored the role of these NPs *in vitro* and *in vivo*. Both model studies show the beneficial response of NPs and warrant further study to confirm the same [26]. Metformin loaded polydopamine nano-formulation promoted the autophagic activity in the PD model by promoting enhancer of zeste homolog 2 mediated proteasomal degradation of phospho-α-Synuclein [26]. Intranasally administered selegiline nanoparticles (SNPs) showed its significant amount in the brain of the PD model. This SNPs intranasal delivery showed a very effective therapeutic response in PD compared to the oral route [27].

CONCLUSION

In conclusion, we can say that different nanoparticles and formulations show better response *via* the intranasal route as compared to others in PD. The major hurdle regarding BBB is solved by this effective route. This nanoparticles based study should be checked in double-blinded clinical trials for PD treatment.

CONSENT FOR PUBLICATION

Not applicable.

CONFLICT OF INTEREST

The authors declare no conflict of interest, financial or otherwise.

ACKNOWLEDGEMENTS

Authors would like to acknowledge UGC Dr. D.S. Kothari Postdoctoral scheme for awarding the fellowship to Dr. Sachchida Nand Rai (Ref. No-F.4-2/2006 (BSR)/BL/19-20/0032).

REFERENCES

[1]   Kelly J, Moyeed R, Carroll C, Albani D, Li X. Gene expression meta-analysis of Parkinson's disease and its relationship with Alzheimer's disease. Mol Brain 2019; 12(1): 16.
        [http://dx.doi.org/10.1186/s13041-019-0436-5] [PMID: 30819229]

[2]   Zhao X, Zhang M, Li C, Jiang X, Su Y, Zhang Y. Benefits of Vitamins in the Treatment of Parkinson's Disease. Oxid Med Cell Longev 2019; 2019: 9426867.
        [http://dx.doi.org/10.1155/2019/9426867] [PMID: 30915197]

[3]   Váradi C. Clinical Features of Parkinson's Disease: The Evolution of Critical Symptoms. Biology (Basel) 2020; 9(5): E103.
        [http://dx.doi.org/10.3390/biology9050103] [PMID: 32438686]

[4]   Zulueta A, Mingione A, Signorelli P, Caretti A, Ghidoni R, Trinchera M. Simple and Complex Sugars in Parkinson's Disease: a Bittersweet Taste. Mol Neurobiol 2020; 57(7): 2934-43.
        [http://dx.doi.org/10.1007/s12035-020-01931-4] [PMID: 32430844]

[5]   Castela I, Hernández LF. Shedding light on dyskinesias. Eur J Neurosci 2020.
        [http://dx.doi.org/10.1111/ejn.14777] [PMID: 32394612]

[6]   Ehlers C, Timpka J, Odin P, Honig H. Levodopa infusion in Parkinson's disease: Individual quality of life. Acta Neurol Scand 2020; 142(3): 248-54.
        [http://dx.doi.org/10.1111/ane.13260] [PMID: 32383152]

[7]   Corona JC. Natural Compounds for the Management of Parkinson's Disease and Attention-Deficit/Hyperactivity Disorder. BioMed Res Int 2018; 2018: 4067597.
        [http://dx.doi.org/10.1155/2018/4067597] [PMID: 30596091]

[8]   Prakash J, Chouhan S, Yadav SK, Westfall S, Rai SN, Singh SP. Withania somnifera alleviates parkinsonian phenotypes by inhibiting apoptotic pathways in dopaminergic neurons. Neurochem Res 2014; 39(12): 2527-36.
        [http://dx.doi.org/10.1007/s11064-014-1443-7] [PMID: 25403619]

[9]   Yadav SK, Rai SN, Singh SP. Mucuna pruriens reduces inducible nitric oxide synthase expression in Parkinsonian mice model. J Chem Neuroanat 2017; 80: 1-10.
[http://dx.doi.org/10.1016/j.jchemneu.2016.11.009] [PMID: 27919828]

[10]   Rai SN, Birla H, Zahra W, Singh SS, Singh SP. Immunomodulation of Parkinson's disease using Mucuna pruriens (Mp). J Chem Neuroanat 2017; 85: 27-35.
[http://dx.doi.org/10.1016/j.jchemneu.2017.06.005] [PMID: 28642128]

[11]   Rai SN, Birla H, Singh SS, *et al. Mucuna pruriens* Protects against MPTP Intoxicated Neuroinflammation in Parkinson's Disease through NF-κB/pAKT Signaling Pathways. Front Aging Neurosci 2017; 9: 421.
[http://dx.doi.org/10.3389/fnagi.2017.00421] [PMID: 29311905]

[12]   Birla H, Rai SN, Singh SS, *et al.* Tinospora cordifolia Suppresses Neuroinflammation in Parkinsonian Mouse Model. Neuromolecular Med 2019; 21(1): 42-53.
[http://dx.doi.org/10.1007/s12017-018-08521-7] [PMID: 30644041]

[13]   Singh SS, Rai SN, Birla H, *et al.* Effect of Chlorogenic Acid Supplementation in MPTP-Intoxicated Mouse. Front Pharmacol 2018; 9: 757.
[http://dx.doi.org/10.3389/fphar.2018.00757] [PMID: 30127737]

[14]   Rai SN, Zahra W, Singh SS, *et al.* Anti-inflammatory Activity of Ursolic Acid in MPTP-Induced Parkinsonian Mouse Model. Neurotox Res 2019; 36(3): 452-62.
[http://dx.doi.org/10.1007/s12640-019-00038-6] [PMID: 31016688]

[15]   Rai SN, Yadav SK, Singh D, Singh SP. Ursolic acid attenuates oxidative stress in nigrostriatal tissue and improves neurobehavioral activity in MPTP-induced Parkinsonian mouse model. J Chem Neuroanat 2016; 71: 41-9.
[http://dx.doi.org/10.1016/j.jchemneu.2015.12.002] [PMID: 26686287]

[16]   Teleanu DM, Chircov C, Grumezescu AM, Volceanov A, Teleanu RI. Blood-Brain Delivery Methods Using Nanotechnology. Pharmaceutics 2018; 10(4): 269.
[http://dx.doi.org/10.3390/pharmaceutics10040269] [PMID: 30544966]

[17]   Singh R, Lillard JW Jr. Nanoparticle-based targeted drug delivery. Exp Mol Pathol 2009; 86(3): 215-23.
[http://dx.doi.org/10.1016/j.yexmp.2008.12.004] [PMID: 19186176]

[18]   Hu J, Gao G, He M, *et al.* Optimal route of gold nanoclusters administration in mice targeting Parkinson's disease. Nanomedicine (Lond) 2020; 15(6): 563-80.
[http://dx.doi.org/10.2217/nnm-2019-0268] [PMID: 32079495]

[19]   Chung TH, Hsu SC, Wu SH, *et al.* Dextran-coated iron oxide nanoparticle-improved therapeutic effects of human mesenchymal stem cells in a mouse model of Parkinson's disease. Nanoscale 2018; 10(6): 2998-3007.
[http://dx.doi.org/10.1039/C7NR06976F] [PMID: 29372743]

[20]   Hegazy MA, Maklad HM, Samy DM, Abdelmonsif DA, El Sabaa BM, Elnozahy FY. Cerium oxide nanoparticles could ameliorate behavioral and neurochemical impairments in 6-hydroxydopamine induced Parkinson's disease in rats. Neurochem Int 2017; 108: 361-71.
[http://dx.doi.org/10.1016/j.neuint.2017.05.011] [PMID: 28527632]

[21]   Smith Y, Wichmann T, Factor SA, DeLong MR. Parkinson's disease therapeutics: new developments and challenges since the introduction of levodopa. Neuropsychopharmacology 2012; 37(1): 213-46.
[http://dx.doi.org/10.1038/npp.2011.212] [PMID: 21956442]

[22]   Nutt JG, Burchiel KJ, Comella CL, *et al.* Randomized, double-blind trial of glial cell line-derived neurotrophic factor (GDNF) in PD. Neurology 2003; 60(1): 69-73.
[http://dx.doi.org/10.1212/WNL.60.1.69] [PMID: 12525720]

[23]   Hernando S, Herran E, Figueiro-Silva J, *et al.* Intranasal Administration of TAT-Conjugated Lipid Nanocarriers Loading GDNF for Parkinson's Disease. Mol Neurobiol 2018; 55(1): 145-55.

[http://dx.doi.org/10.1007/s12035-017-0728-7] [PMID: 28866799]

[24] Hu K, Chen X, Chen W, *et al.* Neuroprotective effect of gold nanoparticles composites in Parkinson's disease model. Nanomedicine (Lond) 2018; 14(4): 1123-36.
[http://dx.doi.org/10.1016/j.nano.2018.01.020] [PMID: 29474924]

[25] Herrán E, Requejo C, Ruiz-Ortega JA, *et al.* Increased antiparkinson efficacy of the combined administration of VEGF- and GDNF-loaded nanospheres in a partial lesion model of Parkinson's disease. Int J Nanomedicine 2014; 9: 2677-87.
[PMID: 24920904]

[26] Bi C, Wang A, Chu Y, *et al.* Intranasal delivery of rotigotine to the brain with lactoferrin-modified PEG-PLGA nanoparticles for Parkinson's disease treatment. Int J Nanomedicine 2016; 11: 6547-59.
[http://dx.doi.org/10.2147/IJN.S120939] [PMID: 27994458]

[27] Sridhar V, Gaud R, Bajaj A, Wairkar S. Pharmacokinetics and pharmacodynamics of intranasally administered selegiline nanoparticles with improved brain delivery in Parkinson's disease. Nanomedicine (Lond) 2018; 14(8): 2609-18.
[http://dx.doi.org/10.1016/j.nano.2018.08.004] [PMID: 30171904]

SUBJECT INDEX

www.ingramcontent.com/pod-product-compliance
Lightning Source LLC
Chambersburg PA
CBHW050836220326
41598CB00006B/380